AF328764

FORMAT AND CONTENT OF THE PACKAGE DESIGN SAFETY REPORT FOR THE TRANSPORT OF RADIOACTIVE MATERIAL

The following States are Members of the International Atomic Energy Agency:

AFGHANISTAN	GERMANY	PALAU
ALBANIA	GHANA	PANAMA
ALGERIA	GREECE	PAPUA NEW GUINEA
ANGOLA	GRENADA	PARAGUAY
ANTIGUA AND BARBUDA	GUATEMALA	PERU
ARGENTINA	GUYANA	PHILIPPINES
ARMENIA	HAITI	POLAND
AUSTRALIA	HOLY SEE	PORTUGAL
AUSTRIA	HONDURAS	QATAR
AZERBAIJAN	HUNGARY	REPUBLIC OF MOLDOVA
BAHAMAS	ICELAND	ROMANIA
BAHRAIN	INDIA	RUSSIAN FEDERATION
BANGLADESH	INDONESIA	RWANDA
BARBADOS	IRAN, ISLAMIC REPUBLIC OF	SAINT KITTS AND NEVIS
BELARUS	IRAQ	SAINT LUCIA
BELGIUM	IRELAND	SAINT VINCENT AND
BELIZE	ISRAEL	THE GRENADINES
BENIN	ITALY	SAMOA
BOLIVIA, PLURINATIONAL	JAMAICA	SAN MARINO
STATE OF	JAPAN	SAUDI ARABIA
BOSNIA AND HERZEGOVINA	JORDAN	SENEGAL
BOTSWANA	KAZAKHSTAN	SERBIA
BRAZIL	KENYA	SEYCHELLES
BRUNEI DARUSSALAM	KOREA, REPUBLIC OF	SIERRA LEONE
BULGARIA	KUWAIT	SINGAPORE
BURKINA FASO	KYRGYZSTAN	SLOVAKIA
BURUNDI	LAO PEOPLE'S DEMOCRATIC	SLOVENIA
CAMBODIA	REPUBLIC	SOUTH AFRICA
CAMEROON	LATVIA	SPAIN
CANADA	LEBANON	SRI LANKA
CENTRAL AFRICAN	LESOTHO	SUDAN
REPUBLIC	LIBERIA	SWEDEN
CHAD	LIBYA	SWITZERLAND
CHILE	LIECHTENSTEIN	SYRIAN ARAB REPUBLIC
CHINA	LITHUANIA	TAJIKISTAN
COLOMBIA	LUXEMBOURG	THAILAND
COMOROS	MADAGASCAR	TOGO
CONGO	MALAWI	TONGA
COSTA RICA	MALAYSIA	TRINIDAD AND TOBAGO
CÔTE D'IVOIRE	MALI	TUNISIA
CROATIA	MALTA	TÜRKİYE
CUBA	MARSHALL ISLANDS	TURKMENISTAN
CYPRUS	MAURITANIA	UGANDA
CZECH REPUBLIC	MAURITIUS	UKRAINE
DEMOCRATIC REPUBLIC	MEXICO	UNITED ARAB EMIRATES
OF THE CONGO	MONACO	UNITED KINGDOM OF
DENMARK	MONGOLIA	GREAT BRITAIN AND
DJIBOUTI	MONTENEGRO	NORTHERN IRELAND
DOMINICA	MOROCCO	UNITED REPUBLIC
DOMINICAN REPUBLIC	MOZAMBIQUE	OF TANZANIA
ECUADOR	MYANMAR	UNITED STATES OF AMERICA
EGYPT	NAMIBIA	URUGUAY
EL SALVADOR	NEPAL	UZBEKISTAN
ERITREA	NETHERLANDS	VANUATU
ESTONIA	NEW ZEALAND	VENEZUELA, BOLIVARIAN
ESWATINI	NICARAGUA	REPUBLIC OF
ETHIOPIA	NIGER	VIET NAM
FIJI	NIGERIA	YEMEN
FINLAND	NORTH MACEDONIA	ZAMBIA
FRANCE	NORWAY	ZIMBABWE
GABON	OMAN	
GEORGIA	PAKISTAN	

The Agency's Statute was approved on 23 October 1956 by the Conference on the Statute of the IAEA held at United Nations Headquarters, New York; it entered into force on 29 July 1957. The Headquarters of the Agency are situated in Vienna. Its principal objective is "to accelerate and enlarge the contribution of atomic energy to peace, health and prosperity throughout the world".

IAEA SAFETY STANDARDS SERIES No. SSG-66

FORMAT AND CONTENT OF THE PACKAGE DESIGN SAFETY REPORT FOR THE TRANSPORT OF RADIOACTIVE MATERIAL

SPECIFIC SAFETY GUIDE

INTERNATIONAL ATOMIC ENERGY AGENCY
VIENNA, 2022

COPYRIGHT NOTICE

All IAEA scientific and technical publications are protected by the terms of the Universal Copyright Convention as adopted in 1952 (Berne) and as revised in 1972 (Paris). The copyright has since been extended by the World Intellectual Property Organization (Geneva) to include electronic and virtual intellectual property. Permission to use whole or parts of texts contained in IAEA publications in printed or electronic form must be obtained and is usually subject to royalty agreements. Proposals for non-commercial reproductions and translations are welcomed and considered on a case-by-case basis. Enquiries should be addressed to the IAEA Publishing Section at:

Marketing and Sales Unit, Publishing Section
International Atomic Energy Agency
Vienna International Centre
PO Box 100
1400 Vienna, Austria
fax: +43 1 26007 22529
tel.: +43 1 2600 22417
email: sales.publications@iaea.org
www.iaea.org/publications

© IAEA, 2022

Printed by the IAEA in Austria
June 2022
STI/PUB/1980

IAEA Library Cataloguing in Publication Data

Names: International Atomic Energy Agency.
Title: Format and content of the package design safety report for the transport of radioactive material / International Atomic Energy Agency.
Description: Vienna : International Atomic Energy Agency, 2022. | Series: IAEA safety standards series, ISSN 1020–525X ; no. SSG-66 | Includes bibliographical references.
Identifiers: IAEAL 21-01468 | ISBN 978-92-0-141321-5 (paperback : alk. paper) | ISBN 978-92-0-141221-8 (pdf) | ISBN 978-92-0-142321-4 (epub)
Subjects: LCSH: Radioactive substances — Packaging. | Radioactive substances — Transportation. | Radioactive substances — Safety regulations.
Classification: UDC 656.073 | STI/PUB/1980

FOREWORD

by Rafael Mariano Grossi
Director General

The IAEA's Statute authorizes it to "establish…standards of safety for protection of health and minimization of danger to life and property". These are standards that the IAEA must apply to its own operations, and that States can apply through their national regulations.

The IAEA started its safety standards programme in 1958 and there have been many developments since. As Director General, I am committed to ensuring that the IAEA maintains and improves upon this integrated, comprehensive and consistent set of up to date, user friendly and fit for purpose safety standards of high quality. Their proper application in the use of nuclear science and technology should offer a high level of protection for people and the environment across the world and provide the confidence necessary to allow for the ongoing use of nuclear technology for the benefit of all.

Safety is a national responsibility underpinned by a number of international conventions. The IAEA safety standards form a basis for these legal instruments and serve as a global reference to help parties meet their obligations. While safety standards are not legally binding on Member States, they are widely applied. They have become an indispensable reference point and a common denominator for the vast majority of Member States that have adopted these standards for use in national regulations to enhance safety in nuclear power generation, research reactors and fuel cycle facilities as well as in nuclear applications in medicine, industry, agriculture and research.

The IAEA safety standards are based on the practical experience of its Member States and produced through international consensus. The involvement of the members of the Safety Standards Committees, the Nuclear Security Guidance Committee and the Commission on Safety Standards is particularly important, and I am grateful to all those who contribute their knowledge and expertise to this endeavour.

The IAEA also uses these safety standards when it assists Member States through its review missions and advisory services. This helps Member States in the application of the standards and enables valuable experience and insight to be shared. Feedback from these missions and services, and lessons identified from events and experience in the use and application of the safety standards, are taken into account during their periodic revision.

I believe the IAEA safety standards and their application make an invaluable contribution to ensuring a high level of safety in the use of nuclear technology. I encourage all Member States to promote and apply these standards, and to work with the IAEA to uphold their quality now and in the future.

THE IAEA SAFETY STANDARDS

BACKGROUND

Radioactivity is a natural phenomenon and natural sources of radiation are features of the environment. Radiation and radioactive substances have many beneficial applications, ranging from power generation to uses in medicine, industry and agriculture. The radiation risks to workers and the public and to the environment that may arise from these applications have to be assessed and, if necessary, controlled.

Activities such as the medical uses of radiation, the operation of nuclear installations, the production, transport and use of radioactive material, and the management of radioactive waste must therefore be subject to standards of safety.

Regulating safety is a national responsibility. However, radiation risks may transcend national borders, and international cooperation serves to promote and enhance safety globally by exchanging experience and by improving capabilities to control hazards, to prevent accidents, to respond to emergencies and to mitigate any harmful consequences.

States have an obligation of diligence and duty of care, and are expected to fulfil their national and international undertakings and obligations.

International safety standards provide support for States in meeting their obligations under general principles of international law, such as those relating to environmental protection. International safety standards also promote and assure confidence in safety and facilitate international commerce and trade.

A global nuclear safety regime is in place and is being continuously improved. IAEA safety standards, which support the implementation of binding international instruments and national safety infrastructures, are a cornerstone of this global regime. The IAEA safety standards constitute a useful tool for contracting parties to assess their performance under these international conventions.

THE IAEA SAFETY STANDARDS

The status of the IAEA safety standards derives from the IAEA's Statute, which authorizes the IAEA to establish or adopt, in consultation and, where appropriate, in collaboration with the competent organs of the United Nations and with the specialized agencies concerned, standards of safety for protection of health and minimization of danger to life and property, and to provide for their application.

With a view to ensuring the protection of people and the environment from harmful effects of ionizing radiation, the IAEA safety standards establish fundamental safety principles, requirements and measures to control the radiation exposure of people and the release of radioactive material to the environment, to restrict the likelihood of events that might lead to a loss of control over a nuclear reactor core, nuclear chain reaction, radioactive source or any other source of radiation, and to mitigate the consequences of such events if they were to occur. The standards apply to facilities and activities that give rise to radiation risks, including nuclear installations, the use of radiation and radioactive sources, the transport of radioactive material and the management of radioactive waste.

Safety measures and security measures[1] have in common the aim of protecting human life and health and the environment. Safety measures and security measures must be designed and implemented in an integrated manner so that security measures do not compromise safety and safety measures do not compromise security.

The IAEA safety standards reflect an international consensus on what constitutes a high level of safety for protecting people and the environment from harmful effects of ionizing radiation. They are issued in the IAEA Safety Standards Series, which has three categories (see Fig. 1).

Safety Fundamentals

Safety Fundamentals present the fundamental safety objective and principles of protection and safety, and provide the basis for the safety requirements.

Safety Requirements

An integrated and consistent set of Safety Requirements establishes the requirements that must be met to ensure the protection of people and the environment, both now and in the future. The requirements are governed by the objective and principles of the Safety Fundamentals. If the requirements are not met, measures must be taken to reach or restore the required level of safety. The format and style of the requirements facilitate their use for the establishment, in a harmonized manner, of a national regulatory framework. Requirements, including numbered 'overarching' requirements, are expressed as 'shall' statements. Many requirements are not addressed to a specific party, the implication being that the appropriate parties are responsible for fulfilling them.

Safety Guides

Safety Guides provide recommendations and guidance on how to comply with the safety requirements, indicating an international consensus that it

---

[1] See also publications issued in the IAEA Nuclear Security Series.

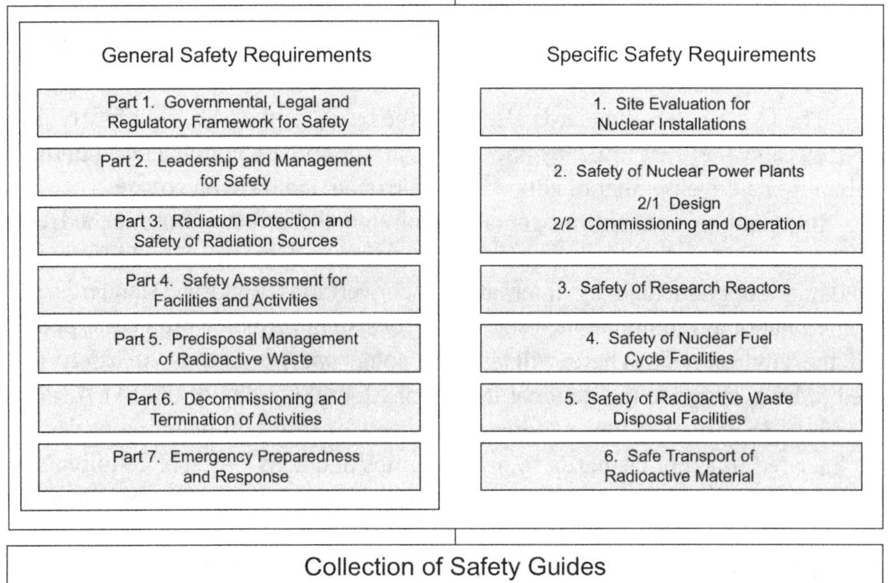

FIG. 1. The long term structure of the IAEA Safety Standards Series.

is necessary to take the measures recommended (or equivalent alternative measures). The Safety Guides present international good practices, and increasingly they reflect best practices, to help users striving to achieve high levels of safety. The recommendations provided in Safety Guides are expressed as 'should' statements.

APPLICATION OF THE IAEA SAFETY STANDARDS

The principal users of safety standards in IAEA Member States are regulatory bodies and other relevant national authorities. The IAEA safety standards are also used by co-sponsoring organizations and by many organizations that design, construct and operate nuclear facilities, as well as organizations involved in the use of radiation and radioactive sources.

The IAEA safety standards are applicable, as relevant, throughout the entire lifetime of all facilities and activities — existing and new — utilized for peaceful purposes and to protective actions to reduce existing radiation risks. They can be used by States as a reference for their national regulations in respect of facilities and activities.

The IAEA's Statute makes the safety standards binding on the IAEA in relation to its own operations and also on States in relation to IAEA assisted operations.

The IAEA safety standards also form the basis for the IAEA's safety review services, and they are used by the IAEA in support of competence building, including the development of educational curricula and training courses.

International conventions contain requirements similar to those in the IAEA safety standards and make them binding on contracting parties. The IAEA safety standards, supplemented by international conventions, industry standards and detailed national requirements, establish a consistent basis for protecting people and the environment. There will also be some special aspects of safety that need to be assessed at the national level. For example, many of the IAEA safety standards, in particular those addressing aspects of safety in planning or design, are intended to apply primarily to new facilities and activities. The requirements established in the IAEA safety standards might not be fully met at some existing facilities that were built to earlier standards. The way in which IAEA safety standards are to be applied to such facilities is a decision for individual States.

The scientific considerations underlying the IAEA safety standards provide an objective basis for decisions concerning safety; however, decision makers must also make informed judgements and must determine how best to balance the benefits of an action or an activity against the associated radiation risks and any other detrimental impacts to which it gives rise.

DEVELOPMENT PROCESS FOR THE IAEA SAFETY STANDARDS

The preparation and review of the safety standards involves the IAEA Secretariat and five Safety Standards Committees, for emergency preparedness and response (EPReSC) (as of 2016), nuclear safety (NUSSC), radiation safety (RASSC), the safety of radioactive waste (WASSC) and the safe transport of radioactive material (TRANSSC), and a Commission on Safety Standards (CSS) which oversees the IAEA safety standards programme (see Fig. 2).

All IAEA Member States may nominate experts for the Safety Standards Committees and may provide comments on draft standards. The membership of the Commission on Safety Standards is appointed by the Director General and includes senior governmental officials having responsibility for establishing national standards.

A management system has been established for the processes of planning, developing, reviewing, revising and establishing the IAEA safety standards.

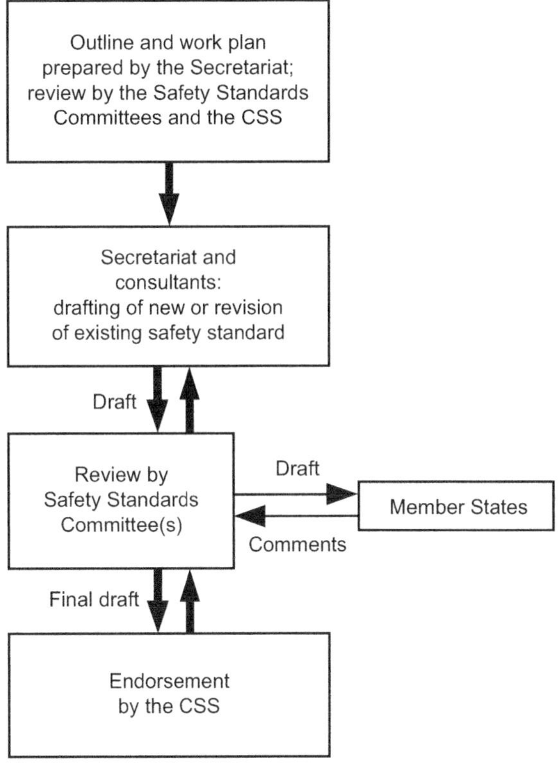

FIG. 2. The process for developing a new safety standard or revising an existing standard.

It articulates the mandate of the IAEA, the vision for the future application of the safety standards, policies and strategies, and corresponding functions and responsibilities.

INTERACTION WITH OTHER INTERNATIONAL ORGANIZATIONS

The findings of the United Nations Scientific Committee on the Effects of Atomic Radiation (UNSCEAR) and the recommendations of international expert bodies, notably the International Commission on Radiological Protection (ICRP), are taken into account in developing the IAEA safety standards. Some safety standards are developed in cooperation with other bodies in the United Nations system or other specialized agencies, including the Food and Agriculture Organization of the United Nations, the United Nations Environment Programme, the International Labour Organization, the OECD Nuclear Energy Agency, the Pan American Health Organization and the World Health Organization.

INTERPRETATION OF THE TEXT

Safety related terms are to be understood as defined in the IAEA Safety Glossary (see https://www.iaea.org/resources/safety-standards/safety-glossary). Otherwise, words are used with the spellings and meanings assigned to them in the latest edition of The Concise Oxford Dictionary. For Safety Guides, the English version of the text is the authoritative version.

The background and context of each standard in the IAEA Safety Standards Series and its objective, scope and structure are explained in Section 1, Introduction, of each publication.

Material for which there is no appropriate place in the body text (e.g. material that is subsidiary to or separate from the body text, is included in support of statements in the body text, or describes methods of calculation, procedures or limits and conditions) may be presented in appendices or annexes.

An appendix, if included, is considered to form an integral part of the safety standard. Material in an appendix has the same status as the body text, and the IAEA assumes authorship of it. Annexes and footnotes to the main text, if included, are used to provide practical examples or additional information or explanation. Annexes and footnotes are not integral parts of the main text. Annex material published by the IAEA is not necessarily issued under its authorship; material under other authorship may be presented in annexes to the safety standards. Extraneous material presented in annexes is excerpted and adapted as necessary to be generally useful.

CONTENTS

1. INTRODUCTION

BACKGROUND

1.1.   Requirements for the safe transport of radioactive material are established in IAEA Safety Standards Series No. SSR-6 (Rev. 1), Regulations for the Safe Transport of Radioactive Material, 2018 Edition [1] (hereinafter referred to as the 'Transport Regulations').

1.2.   Packages intended for the transport of radioactive material are required to be designed to meet the applicable national and international regulations, and, as such, documentary evidence of compliance of a package design with the applicable regulations is required.

1.3.   For package designs that require approval by a competent authority, the package design safety report (PDSR) is the basis for the application to the competent authority for approval of the design.

1.4.   For package designs where competent authority approval is not required, documentary evidence of compliance of the package design with all applicable requirements is required to meet para. 801 of the Transport Regulations. In addition, for packages that do not require competent authority approval, some form of 'certificate of compliance' with all the applicable requirements of the Transport Regulations might need to be applied (see para. 801.3 of IAEA Safety Standards Series No. SSG-26 (Rev. 1), Advisory Material for the IAEA Regulations for the Safe Transport of Radioactive Material (2018 Edition) [2]), and a PDSR would be an appropriate form of documentary evidence of compliance with the Transport Regulations.

1.5.   In this Safety Guide, all documentary evidence of compliance of a package design with the Transport Regulations will be called a PDSR, irrespective of whether the package design requires competent authority approval.

1.6.   This Safety Guide is based on the Transport Regulations. The Transport Regulations also constitute the basis of the United Nations Recommendations on the Transport of Dangerous Goods [3] and of other relevant international, regional and national regulations, for all modes of transport.

1.7.   The format of this Safety Guide is based on the European Technical Guide on Package Design Safety Reports for the Transport of Radioactive Material [4].

1.8. The definitions stated in the Transport Regulations and the terms defined and explained in the IAEA Safety Glossary [5] apply to this Safety Guide.

OBJECTIVE

1.9. The objective of this Safety Guide is to provide recommendations on the preparation of a PDSR to demonstrate compliance of a package design for the transport of radioactive material with the Transport Regulations.

1.10. This Safety Guide is intended for use by applicants for approval of package designs (when package designs are subject to competent authority approval) as well as by package designers and/or consignors (when package designs do not require competent authority approval) to demonstrate compliance with the requirements of the Transport Regulations applicable to the respective package type.

1.11. This Safety Guide could be also used to demonstrate compliance with any national or international regulations relating to package design, if the national or international regulations are based on the Transport Regulations. This Safety Guide does not replace the Transport Regulations or limit their application. Moreover, following the recommendations of this Safety Guide does not relieve the package designer of the need for any additional analysis associated with a specific package design as required by the competent authority.

1.12. This Safety Guide provides an example of the structure and format for a PDSR but does not intend to replace any existing format for a PDSR that might be determined by national regulations or standards for packages intended for domestic use only.

1.13. In the event of a conflict or anomaly between the provisions of the Transport Regulations and this Safety Guide, the requirements in the Transport Regulations apply. For regulatory purposes, reference should be made to the detailed provisions of the Transport Regulations.

SCOPE

1.14. This Safety Guide covers package designs requiring competent authority approval in accordance with para. 802 of the Transport Regulations for the following types of package:

— Type B(U) packages;
— Type B(M) packages;
— Type C packages;
— Packages containing fissile material not excepted by paras 417, 674 or 675 of the Transport Regulations;
— Packages designed to contain 0.1 kg or more of uranium hexafluoride.

1.15. This Safety Guide also covers package designs not requiring competent authority approval for the following types of package:

— Excepted packages;
— Industrial packages Type 1 (Type IP-1);
— Industrial packages Type 2 (Type IP-2);
— Industrial packages Type 3 (Type IP-3);
— Type A packages.

1.16. Designs for special form radioactive material, designs for low dispersible radioactive material, unpackaged low specific activity material (LSA-I material) and unpackaged surface contaminated objects (SCO-I and SCO-III) are outside the scope of this Safety Guide.

STRUCTURE

1.17. Section 2 of this Safety Guide provides general recommendations on the PDSR. Appendices I–IV describe the information that should be provided in the PDSR for, respectively, excepted packages; industrial packages; Type A packages; and Type B(U), Type B(M) and Type C packages. Appendix V describes the additional information that should be provided in the PDSR for packages containing fissile material. Appendix VI describes the additional information that should be provided for packages containing 0.1 kg or more of uranium hexafluoride. Annex I provides a matrix of the applicable paragraphs of the Transport Regulations to be included in the demonstration of compliance for each package type and additional provisions for packages containing fissile nuclides and packages containing 0.1 kg or more of uranium hexafluoride. Annex II provides a list of reference publications used by different competent authorities for the technical assessment of package designs. Annex III provides information on the structure of the PDSR for the packages addressed in Appendices I–IV.

2. GENERAL RECOMMENDATIONS FOR THE PACKAGE DESIGN SAFETY REPORT

2.1. The package designer is the person or organization that takes responsibility for the complete package design. For each package design there should be only one package designer, who should also issue the PDSR.

2.2. In accordance with para. 306 of the Transport Regulations, a management system is required for package design. The PDSR should be a controlled document[1], approved for issue in accordance with the package designer's management system. It should be signed and dated and should indicate its revision number or issue status. The PDSR should include a contents list, and the total number of pages and annexes of the PDSR should be indicated. Any changes between revisions of the PDSR should be clearly documented. The PDSR should include a record of its production, review and approval by the package designer. Further recommendations on the management system for transport are provided in IAEA Safety Standards Series No. TS-G-1.4, The Management System for the Safe Transport of Radioactive Material [6]. A PDSR is typically divided into two parts. Part 1 of the PDSR should contain the information that supports the demonstration of compliance with the Transport Regulations. Part 2 of the PDSR should contain detailed technical analyses supporting this information. Figure 1 shows a generic structure and the recommended contents of a PDSR that apply to all package types.

---

[1] A 'controlled document' is a document that is approved and maintained within a management system.

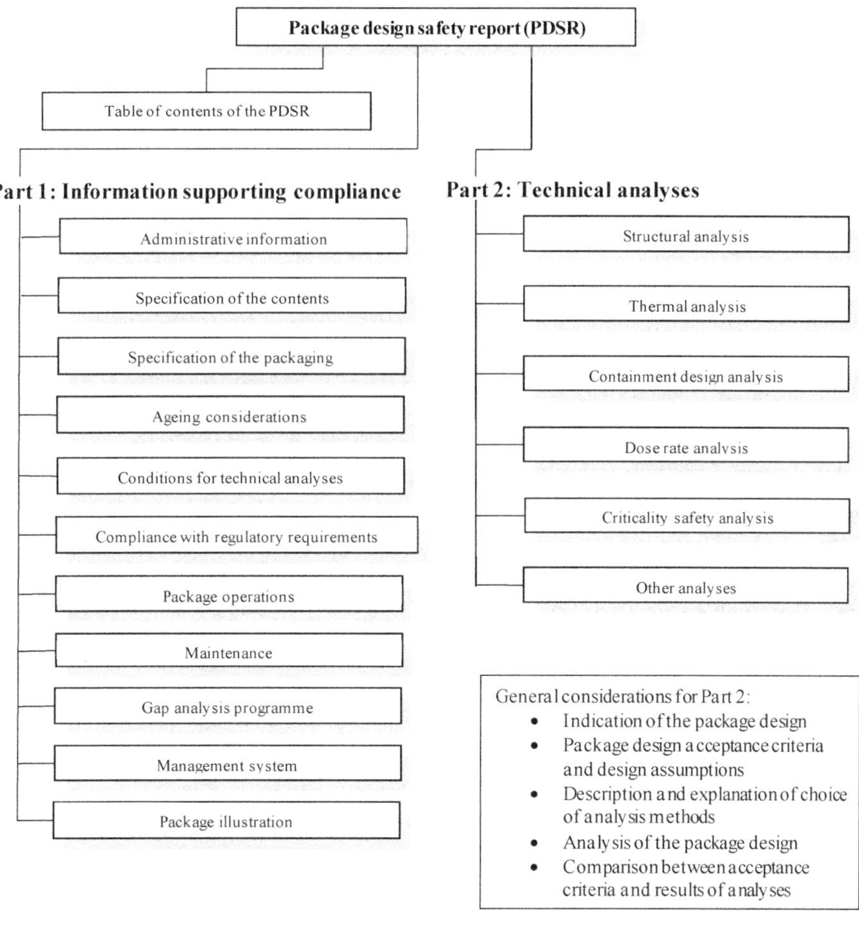

FIG. 1. Structure and contents of a package design safety report.

2.3.   The structure of the PDSR should be the following, as shown in Fig. 1:

— Table of contents of the PDSR.
— Part 1: Information supporting the demonstration of compliance with the Transport Regulations:
  • Administrative information;
  • Specification of the contents;
  • Specification of the packaging;
  • Ageing considerations;
  • Conditions for technical analyses;
  • Compliance with regulatory requirements;

- Package operations;
- Maintenance;
- Gap analysis programme;
- Management system;
- Package illustration.

— Part 2: Technical analyses:

- Structural analysis;
- Thermal analysis;
- Containment design analysis;
- Dose rate analysis;
- Criticality safety analysis;
- Other analyses.

2.4. A table showing the structure of the PDSR for Appendices I–VI is provided in Annex III.

2.5. The following general considerations should be taken into account for all technical analyses in Part 2 of the PDSR:

(a) Indication of the package design;
(b) Package design acceptance criteria[2] and design assumptions;
(c) Description and explanation of the choice of analysis methods;
(d) Analysis of the package design;
(e) Comparison between the acceptance criteria and the results of the analyses.

2.6. The PDSR may be compiled as a single document or as an integrated set of individual documents. Each individual document in Part 1 and Part 2 of the PDSR should also be a controlled document. Each individual document in Part 2 of the PDSR should be produced and verified by specialists in the technical area being assessed.

2.7. The scope and technical content of the PDSR should be set at the appropriate level for each type of package by adopting a graded approach consistent with para. 104 of the Transport Regulations. Depending on the package type, some of the items included in the contents of the PDSR in para. 2.3 are not needed.

---

[2] In the context of this Safety Guide, 'acceptance criteria' are the limits on the value of an indicator (i.e.. loss of radioactive content dose rate, temperature, subcriticality,) specified by the regulatory body and used to assess the ability of a package to perform its function according to the design.

More detailed information on the content of the PDSR for each package type is provided in Appendices I–VI.

2.8. A PDSR should contain controlled engineering drawings that show the details used in the demonstration of compliance with the regulations. For complex packages, this may involve several large drawings, necessary for modelling and assessing the package for calculations relating to mechanical strength, heat transfer, dose rates and criticality; these are often called 'design drawings'. For simple packages, such as excepted packages, that do not need the same level of detail as complex packages for the demonstration of compliance, these drawings may be simple and are often called 'schematic drawings'.

2.9. A reproducible illustration of the package should be included at the end of Part 1 of the PDSR. In accordance with para 809(i) of the Transport Regulations, a reproducible illustration, not larger than 21 cm × 30 cm, showing the make-up of the package is requested to be part of the application for approval. This illustration does not need to be very detailed but should present an overview of the package and should fit on one page.

2.10. The International System of Units (SI) should be used throughout the PDSR.

Appendix I

EXCEPTED PACKAGES

I.1. This appendix provides specific recommendations on the information that should be included in Parts 1 and 2 of the PDSR for excepted packages. Table 1 lists each item of the PDSR, with applicable information and guidance. Further recommendations on excepted packages are provided in SSG-26 (Rev. 1) [2].

I.2. In accordance with para. 801 of the Transport Regulations, excepted packages require a PDSR. In accordance with the graded approach and taking into consideration the lower risks presented by excepted packages, their PDSRs may be less extensive than those for other types of package.

I.3. For packages containing fissile materials to be transported as excepted packages, one of the provisions of para. 417 of the Transport Regulations is required to apply.

I.4. Packages containing less than 0.1 kg of uranium hexafluoride may be classified as UN 3507 and transported as excepted packages under the provisions of para. 425 of the Transport Regulations. More information on the applicable regulations for the carriage of UN 3507 is available in chapter 3.2 of Ref. [3].

I.5. For technical analyses, compliance of the package with the Transport Regulations should be demonstrated by the package designer either directly in item 1.4 of the PDSR or, if necessary, in Part 2 of the PDSR.

I.6. For each of the technical analyses in Part 2 of the PDSR, the following considerations apply:

(a) The package design being evaluated should be uniquely identified by precisely indicating a schematic drawing of the packaging (see item 1.3 of the PDSR), including its revision number, and the specification of the contents (see item 1.2 of the PDSR), including its revision number.
(b) The acceptance criteria for the technical analyses and the package design assumptions relating to geometry and performance characteristics should be defined and justified when necessary. The acceptance criteria should be specified by the designer and should be derived from the criteria established by the regulatory body and from other applicable standards developed to meet the regulatory requirements. The design assumptions refer to the design

specifications provided in items 1.2 and 1.3 of the PDSR or to other assumptions derived from the design specifications and used in the technical analyses.

(c)   The results of the technical analyses should be compared with the acceptance criteria and package design assumptions, and regulatory compliance should be justified accordingly.

I.7.   The following items listed in Fig. 1 and para. 2.3 of this Safety Guide are not relevant to and not needed for excepted packages and, therefore, are not included in Table 1:

— Part 1:
  - Ageing considerations;
  - Conditions for technical analyses;
  - Gap analysis programme;
  - Package illustration.
— Part 2:
  - Thermal analysis;
  - Containment design analysis;
  - Criticality safety analysis;
  - Other analyses.

APPENDIX I
TABLE 1. PACKAGE DESIGN SAFETY REPORT FOR
EXCEPTED PACKAGES

Table of contents of the package design safety report (PDSR)

The contents of the PDSR, Part 1 and Part 2, should be listed here, including the revision number of each individual document included in the PDSR.

Part 1

1.1. ADMINISTRATIVE INFORMATION

The following administrative information should be provided:

(a)   Identification of the package designer (i.e. name, address, contact details);
(b)   UN number of the package, as applicable;

(c)   Modes of transport for which the package is designed, and any operational restrictions associated with any mode of transport;

(d)   Reference to the applicable regulations for the specific package design, including the edition of the Transport Regulations to which the package design is referring.

1.2. SPECIFICATION OF THE CONTENTS

A detailed description of the permitted contents of the package should be provided, stating — at a minimum — the following information, as applicable:

(a)   The general nature of contents (e.g. articles, instruments, metallurgical specimens, internal contamination of the package).

(b)   The nuclides and/or nuclide composition, including progeny radionuclides.

(c)   The A_1 and/or A_2 values of the radionuclides to be carried in the package. The A_1 and/or A_2 values for radionuclides that are not listed in table 2 of the Transport Regulations are required to be determined in accordance with paras 403–407 of the Transport Regulations and may be subject to multilateral approval in accordance with para. 403 of the Transport Regulations.

(d)   The physical and chemical state of the contents.

(e)   The type and characteristics of the radiation emitted by the contents of the package.

(f)   Limitations on activity, mass and activity concentrations, and heterogeneities in the distribution of the nuclides. Compliance with the activity limits for excepted packages in accordance with table 4, paras 423(e) and 424(c) (for transport by post), para. 425 (for uranium hexafluoride) and para. 427 (for empty packagings) of the Transport Regulations, as applicable, is required.

(g)   A valid certificate for special form radioactive material or low dispersible radioactive material, when such material is included in the package.

(h)   The mass of fissile material, the nuclides and the enrichment of the contents, when fissile material excepted under para. 417 of the Transport Regulations is included in the package.

(i)   Other dangerous properties of the contents. In accordance with para. 618 of the Transport Regulations, any other dangerous properties (subsidiary hazards) of the contents of the package are required to be taken into account in the package design to be in compliance with the relevant transport regulations for dangerous goods. Additional information on the classification and design requirements for dangerous goods in accordance with the predominant subsidiary hazard can be found in chapter 3.3, Special Provision 290, of Ref. [3]. Regarding packages containing less than 0.1 kg of uranium hexafluoride that are excepted in accordance with para. 425 of the Transport Regulations and that are classified as UN 3507, additional information can be found in chapter 3.2 of Ref. [3] (or in the applicable regulations for the carriage of dangerous goods).

(j)   Other limitations on the contents.

1.3. SPECIFICATION OF THE PACKAGING

The packaging design should be defined to the extent necessary to demonstrate compliance with the Transport Regulations; the following information should be included, as applicable:

(a)   Schematic drawings;
(b)   The overall dimensions and the maximum mass of the package when fully loaded;
(c)   A list of all packaging components important to safety and their materials;
(d)   The maximum normal operating pressure (in the case of air transport).

For radioactive material that has other dangerous properties, the packing instructions appropriate to the other dangerous properties of the radioactive material are specified in para. 4.1.9.1.5 of Ref. [3] (or in the applicable regulations for the carriage of dangerous goods). For packages containing less than 0.1 kg of uranium hexafluoride, the packing instruction P603 in chapter 4.1 of Ref. [3] is applicable.

1.4. COMPLIANCE WITH REGULATORY REQUIREMENTS

The PDSR should include a complete list of all paragraphs of the Transport Regulations and other international or national regulations applicable to the package design.

Compliance of the package design with these regulations should be demonstrated in the PDSR. This could be demonstrated using a table (or any other written format) linking the appropriate items of the PDSR, where compliance is demonstrated, to the applicable paragraphs of the regulations.

The applicable paragraphs of the Transport Regulations for excepted packages are provided in a matrix in Annex I.

1.5. PACKAGE OPERATIONS

The minimum specifications for the following activities should be fully defined, as applicable:

(a)   Assembly of the packaging components;
(b)   Loading and unloading of the package contents;
(c)   Controls before each shipment;
(d)   Handling and tie-down.

If written procedures with a detailed description of these activities are available, then reference should be made to these procedures.

1.6. MAINTENANCE

The minimum specifications for the following activities should be fully defined, as applicable:

(a)   Maintenance and inspection before each shipment;
(b)   Maintenance and inspection at periodic intervals throughout the lifetime use of the packaging and/or package.

When developing the maintenance specifications to be included in this item of the PDSR, the following points should be taken into account:

(1)   Inspection of the package before shipment might be sufficient for excepted packages.
(2)   For single use packages, periodic maintenance does not need to be considered.
(3)   If written procedures with details of the maintenance activities are available, then reference should be made to these procedures.

1.7. MANAGEMENT SYSTEM

The PDSR should specify the management system established and implemented by the package designer, in accordance with para. 306 of the Transport Regulations, to demonstrate compliance with the relevant provisions of the Transport Regulations.

The management system should be commensurate with the complexity of the package design and should include a reliable document control system.

Further recommendations on the management system for transport are provided in TS-G-1.4 [6].

Part 2

2.1. STRUCTURAL ANALYSIS

The results of the assessment of mechanical behaviour (including, as applicable, analysis of fatigue, brittle fracture and creep) for routine conditions of transport should be included in this item of the PDSR.

The assessment of mechanical behaviour should demonstrate compliance with the following requirements:

(a) The requirements for the lifting attachments established in paras 608 and 609 of the Transport Regulations;
(b) The requirements for the packaging attachments used to restrain the package within its conveyance during transport;
(c) For packages to be transported by air, the additional requirements for the package components of the containment system established in paras 619–621 of the Transport Regulations.

If the package is to be transported by air, the structural analysis of the containment system should take into account ambient temperatures and pressures likely to be encountered in routine conditions of transport, as well as the specific temperature and pressure requirements for air transport.

Attention should be paid to ensuring that nuts, bolts and other retention devices maintain their safety functions during routine conditions of transport, even after repeated use.

Further recommendations are provided in paras 607.1–621.3 of SSG-26 (Rev. 1) [2].

2.2. DOSE RATE ANALYSIS

The dose rates at the external surface of the package for routine conditions of transport should be assessed to demonstrate compliance with the requirements for excepted packages established in paras 516 and 423(a), as applicable, of the Transport Regulations.

The dose rate analysis should be based on assuming the maximum radioactive content of the package or such content for an excepted package as would create the maximum dose rate at the surface of the package.

The dose rate analysis should take into account the most recent International Commission on Radiological Protection recommendations on nuclear decay data for dosimetric calculations (e.g. see Ref. [7]).

As para. 516.5 of SSG-26 (Rev. 1) [2] states:

> "The maximum dose rate should be determined taking into account potentially significant amplifying phenomena such as movement of the radioactive contents, or, in the case of packages containing liquids, change in the state of the contents, including segregation and/or precipitation of the radionuclides."

If dose rate measurements are used in the analysis, the source used for the measurements should be representative of the radioactive contents specified in the package design.

Appendix II

INDUSTRIAL PACKAGES

II.1. This appendix provides specific recommendations on the information that should be included in Parts 1 and 2 of the PDSR for industrial packages. Table 2 lists each item of the PDSR, with applicable information and guidance. Further recommendations on industrial packages are provided in SSG-26 (Rev. 1) [2].

II.2. In accordance with para. 801 of the Transport Regulations, Type IP-1, Type IP-2 and Type IP-3 packages require a PDSR. In accordance with the graded approach, and taking into consideration the lower risks presented by Type IP-1 packages, the PDSR for Type IP-1 packages may be less developed than those for other types of package.

II.3. For industrial packages containing fissile nuclides, in addition to the recommendations in this appendix, see also Appendix V.

II.4. For packages containing 0.1 kg or more of uranium hexafluoride, in addition to the recommendations in this appendix, see also Appendix VI.

II.5. For technical analyses, compliance of the package with the Transport Regulations should be demonstrated by the package designer either directly in item 1.5 of the PDSR or, if necessary, in Part 2 of the PDSR.

II.6. For each of the technical analyses in Part 2 of the PDSR, the following considerations apply:

(a) The package design being evaluated should be uniquely identified by precisely indicating a drawing of the packaging (see item 1.3 of the PDSR), including its revision number, and the specification of the contents (see item 1.2 of the PDSR), including its revision number.

(b) The acceptance criteria for the technical analyses and the package design assumptions relating to geometry or performance characteristics should be defined and justified when necessary. The acceptance criteria should be specified by the designer and should be derived from the criteria established by the regulatory body and from other applicable standards developed to meet the regulatory requirements. The design assumptions refer to the design specifications provided in items 1.2 and 1.3 of the PDSR or to other assumptions derived from the design specifications and used in the technical analyses. All mechanical, thermal and shielding characteristics of each component of the package and the acceptance criteria used in technical analyses should be defined.

(c)   The demonstration of compliance of a Type IP-2 or Type IP-3 package design is required to be accomplished in accordance with para. 701 of the Transport Regulations by any of the following methods or by a combination thereof:

    (i)   The results of physical testing of prototypes or models of appropriate scale.

    (ii)   Reference to previous satisfactory demonstrations of a sufficiently similar nature. Test results of designs similar to the design under consideration are permissible if the similarity can be demonstrated sufficiently by justification and validation.

    (iii)   Calculation or reasoned argument, when the calculation procedures are generally agreed to be suitable and conservative. Assumptions should be clearly stated and fully justified, including by physical testing if applicable.

Further recommendations are provided in paras 701.1–701.25 of SSG-26 (Rev. 1) [2]. No regulatory testing is required for Type IP-1 packages.

The methods or the standards used in each analysis specified in items 2.1–2.5 of the PDSR should include a description of the analysis technique used, the limitations and accuracy of this technique and a demonstration of the correct application of the technique for the analysis of the package design.

If computer codes are used for the safety analysis, additional information should be included in the PDSR to demonstrate that the code is verified and validated for the specific field of use. The justification for the applicability of these codes should include a statement of possible sources of error and/or uncertainty relative to the effects of the operating platform (computer) used and of modelling assumptions and simplifications, as well as of any other parameter influencing the calculated results and a sensitivity analysis.

(d)   The performance characteristics of the package design should be assessed. More than one sequence of tests might need to be considered to ensure that the various safety functions to be fulfilled by different components of the package design comply with the regulatory requirements. Other hazards (e.g. corrosion, combustion, pyrophoricity or other chemical reactions, radiolysis, phase changes) should be analysed, as necessary, if they have a consequential effect on the safety functions of the package.

(e) The results of the technical analyses should be compared with the acceptance criteria, and package design assumptions and regulatory compliance should be justified accordingly.

II.7. The following items listed in Fig. 1 and para. 2.3 of this Safety Guide are not relevant to and not needed for industrial packages and, therefore, are not included in Table 2:

— Part 1
   • Conditions for technical analyses;
   • Package illustration.
— Part 2
   • Other analyses.

APPENDIX II
TABLE 2. PACKAGE DESIGN SAFETY REPORT FOR
INDUSTRIAL PACKAGES

Table of contents of the package design safety report (PDSR)

The contents of the PDSR, Part 1 and Part 2, should be listed here, including the revision number of each individual document included in the PDSR.

Part 1

1.1. ADMINISTRATIVE INFORMATION

The following administrative information should be provided:

(a) Colloquial name of the package, if applicable;
(b) Identification of the package designer (e.g. name, address, contact details);
(c) Type of industrial package (i.e. Type IP-1, Type IP-2 or Type IP-3);
(d) For package types IP-2 and IP-3, packaging and/or package design identification and restrictions on packaging serial number(s), if applicable;
(e) Modes of transport for which the package is designed, and any operational restrictions associated with any mode of transport;

(f)   Reference to applicable regulations for the specific package design, including the edition of the Transport Regulations to which the package design is referring.

1.2. SPECIFICATION OF THE CONTENTS

A detailed description of the permitted contents of the package design should be provided, stating — at a minimum — the following information, as applicable:

(a)   The general nature of contents (e.g. fresh fuel, contaminated tools, waste).

(b)   The nuclides and/or nuclide composition, including progeny radionuclides.

(c)   The A_1 and A_2 values of the radionuclides to be carried in the package. The A_1 and A_2 values for radionuclides that are not listed in table 2 of the Transport Regulations are required to be determined in accordance with paras 403–407 of the Transport Regulations and may be subject to multilateral approval in accordance with para. 403 of the Transport Regulations.

(d)   The physical and chemical state, geometric shape, arrangement, and material specifications. For industrial packages, the limits of the contents depend on the physical state of the radioactive content.

(e)   The type and characteristics of the radiation emitted by the contents of the package.

(f)   Limitations on activity, mass and activity concentrations, and heterogeneities in the distribution of the nuclides. The contents should be classified appropriately in one of the categories of low specific activity (LSA) material or surface contaminated object (SCO), as appropriate, in accordance with paras 409 and 413 of the Transport Regulations. Limitations on specific activity (Bq/g) and surface contamination (Bq/cm^2) may be required. Conveyance activity limits in accordance with table 6 of the Transport Regulations should also be taken into account to limit the activity of a single package, if applicable.

(g)   The mass of fissile material, the nuclides and the enrichment of the contents (see also Appendix V).

(h)   Other dangerous properties of the contents. In accordance with para. 618 of the Transport Regulations, any other dangerous properties (subsidiary hazards) of the contents of the package are required to be taken into account in the package design to be in compliance with the relevant transport regulations for dangerous goods. Additional information on the design requirements for dangerous goods in accordance with the subsidiary hazard can be found in chapter 3.3, Special Provision 172, of Ref. [3].

(i)   Other limitations on the contents (e.g. moisture, presence of acid). Safety relevant limits for non-radioactive materials — for example, limits in terms of material composition, density, form, location within the package, or restrictions on relative quantities of materials — should be stated.

1.3. SPECIFICATION OF THE PACKAGING

The packaging design should be defined to the extent necessary to demonstrate compliance with the Transport Regulations; the following information should be included, as applicable:

(a)   Design drawings.
(b)   The overall dimensions, the maximum mass of the package when fully loaded, and the mass of the empty packaging.
(c)   A list of all packaging components important to safety and their materials. For Type IP-2 and Type IP-3 packages, the material specifications of the packaging components should also be included.
(d)   The maximum normal operating pressure (particularly in the case of air transport).

For Type IP-3 packages, the specification of the packaging should include a description of the following:

(1)   The packaging body, lid (e.g. closure mechanism and tamper indicating features), internal arrangements, and components for lifting and tie-down.
(2)   The protection against corrosion.
(3)   The protection against contamination.
(4)   The packaging components required for shielding.
(5)   The shock absorbing components.
(6)   Testing specifications and controls before first use to transport radioactive material. This ensures compliance of the fabrication with the design and allows acceptance of the specimen before first use. See also para. 501 of the Transport Regulations.

For radioactive material that has other dangerous properties, see para. 4.1.9.1.5 of Ref. [3]. The packing instructions or tank instructions as specified in Ref. [3] (or in the applicable regulations for the carriage of dangerous goods) appropriate to the other dangerous properties of the radioactive material should be complied with, with the exception of industrial packages containing 0.1 kg or more of uranium hexafluoride.

1.4. AGEING CONSIDERATIONS

Depending on the package design, the information relating to ageing considerations can also be provided by the package designer directly in the table mentioned in item 1.5.

For packaging used once for a single transport and not intended for shipment after storage, this item of the PDSR should be left blank.

For all other packaging, this item of the PDSR should include the following information:

(a) The intended conditions of use of the package that might influence ageing;
(b) The potential ageing mechanisms relevant to the package design, taking into account the intended conditions of use of the package;
(c) Operational measures (including maintenance and inspection activities before shipment) to monitor and limit the ageing effects;
(d) Analysis of the influence of ageing of the packaging and contents on the design assumptions for demonstration of compliance with the regulations, including the technical analyses in Part 2 of the PDSR, considering the specified intended use conditions, ageing mechanisms and operational measures.

Further recommendations are provided in paras 613A.1–613A.5 of SSG-26 (Rev. 1) [2].

1.5. COMPLIANCE WITH REGULATORY REQUIREMENTS

The PDSR should include a complete list of all paragraphs of the Transport Regulations and other international or national regulations applicable to the package design.

Compliance of the package design with these regulations should be demonstrated in the PDSR. This could be done using a table (or any other written format) linking the appropriate items of the PDSR, where compliance is demonstrated, to the applicable paragraphs of the regulations.

The applicable paragraphs of the Transport Regulations for industrial packages are provided in a matrix in Annex I.

1.6. PACKAGE OPERATIONS

The minimum specifications for the following activities should be fully defined, as applicable:

(a) Assembly of the packaging components. This information should not be included for Type IP-1 packages. Type IP-3 packages are required to comply with para. 637 of the Transport Regulations.
(b) Loading and unloading of the package contents.
(c) Testing and controls before each shipment.
(d) Handling and tie-down. Specifications on bolt torquing and number of transport cycles (to be used in fatigue analysis) for each mode of transport should be included, if applicable.
(e) Any proposed supplementary equipment and operational controls to be applied during transport and, if applicable, during storage before transport, including those that might influence ageing mechanisms.

In addition to the radioactive properties, any other dangerous properties of the contents of the package are required to be taken into account (see para. 507 of the Transport Regulations).

If written procedures with a detailed description of these activities are available, then reference should be made to these procedures.

1.7. MAINTENANCE

The minimum specifications for the following activities should be fully defined, as applicable:

(a) Maintenance and inspection before each shipment;
(b) Maintenance and inspection at periodic intervals throughout the lifetime use of the packaging and/or package.

When developing the maintenance specifications to be included in this item of the PDSR, the following points should be taken into account:

(1) Ageing mechanisms during storage should be considered, when applicable.
(2) For single use packages, periodic maintenance does not need to be considered.
(3) If written procedures with details of the maintenance activities are available, then reference should be made to these procedures.

1.8. GAP ANALYSIS PROGRAMME

For packages that are to be used for shipment after storage, the PDSR should include a gap analysis programme describing a systematic procedure for periodic evaluation of changes of regulations, changes in technical knowledge and changes in the state of the package design during storage (see also paras 613A.5, 809.3 and 809.4 of SSG-26 (Rev. 1) [2] and Ref. [8]).

1.9. MANAGEMENT SYSTEM

The PDSR should specify the management system established and implemented by the package designer, in accordance with para. 306 of the Transport Regulations, to demonstrate compliance with the relevant provisions of the Transport Regulations.

The management system should be commensurate with the complexity of the package design and should include a reliable document control system.

Further recommendations on the management system for transport are provided in TS-G-1.4 [6].

Part 2

2.1. STRUCTURAL ANALYSIS

The assessment of the mechanical behaviour (including, as applicable, analysis of fatigue, brittle fracture and creep) for routine conditions of transport for all industrial packages and for normal conditions of transport for Type IP-2 and Type IP-3 packages should include the following:

(a) The components of the containment system. This is required for Type IP-1 packages only when transported by air (see paras 619–621 of the Transport Regulations).
(b) The package components that provide radiation shielding. This is not required for Type IP-1 packages.
(c) Any other package components (e.g. shock absorbing components) whose performance may have a consequential effect on (a) and (b) above. This is not applicable to Type IP-1 packages.
(d) The lifting attachments (see paras 608 and 609 of the Transport Regulations).

(e)   The packaging attachments used to restrain the package within its conveyance for routine conditions of transport.

If the package is to be transported by air (see paras 619–621 of the Transport Regulations), the structural analysis of the containment system should take into account ambient temperatures and pressures likely to be encountered in routine conditions of transport, as well as the specific temperature and pressure requirements for air transport. Further recommendations are provided in paras 621.2 and 621.3 of SSG-26 (Rev. 1) [2].

Attention should be paid to ensuring that nuts, bolts and other retention devices maintain their safety functions during routine conditions of transport, even after repeated use. Further recommendations are provided in para. 613.1 of SSG-26 (Rev. 1) [2].

When performing the structural analysis for industrial packages, except for a Type IP-2 or a Type IP-3 package meeting the alternative requirements in paras 626–630 of the Transport Regulations, the following points should be considered:

(1)   General considerations:
   (i)   The mechanical properties of the materials considered in the safety demonstration should be representative of the range of mechanical properties of the package components, with account taken of the temperatures likely to be encountered during routine conditions of transport (see para. 616 of the Transport Regulations).
   (ii)   The impacts on the package behaviour due to variations in the shock absorbing properties of the shock absorber material (e.g. wood, polymers, plaster, concrete) with temperatures likely to be encountered during routine conditions of transport should be analysed.
   (iii) The safety against brittle fracture at temperatures likely to be encountered during routine conditions of transport of components of the containment system made of potentially brittle materials (e.g. ferritic steels, cast iron) should be analysed, if necessary.
   (iv)   The strength of lid bolts should be verified for all drop orientations.
   (v)   The internal components (e.g. content, basket, cage) should be assessed to verify that they are not liable to damage the containment system.
   (vi)   The condition of the containment system should be assessed to demonstrate compliance with the specifications in item 2.3 of the PDSR for the temperature range likely to be encountered during routine conditions of transport.

      (vii) Phenomena such as radiolysis, internal pressure elevation, internal inflammation or explosion, physical changes and chemical reactions should be considered when analysing the maximum pressure.

(2) Considerations for experimental mechanical testing:

      (i) The package orientations should be determined in accordance with para. 722.4 of SSG-26 (Rev. 1) [2]. The orientations should maximize the loading of the package (in terms of stress, strain, acceleration and deformation), with consideration of the different package components (e.g. cask body, lid system, shock absorber) and of the protection objectives (i.e. containment and shielding).

      (ii) For reduced scale models, geometry and material properties similar to the original design, or conservative geometry and material properties, should be used.

      (iii) The results of the drop test with reduced scale models should be assessed to guarantee that they cover, or are transferable to, the original design.

      (iv) The representativeness of drop tests performed with reduced scale models should be demonstrated.

      (v) The experimental mechanical tests should be conducted and reported in accordance with the management system. The test report should address the verification of the package before testing, the description of the test site, the equipment used for the measurements and its calibration data, and the results of the measurements performed. This report should also contain pictures showing and explaining the conditions under which the tests were performed and their results.

(3) Considerations for calculations:

      (i) See point (2)(i) above.

      (ii) Validated computer codes should be used. Input parameters (e.g. material laws, characteristic values, boundary conditions) should describe sufficiently and precisely the real technical and/or physical problems, and the use of these parameters should be justified.

      (iii) If uncertainties exist regarding important input parameters (e.g. material laws), conservative design calculations — including the possible range of material properties — should be performed.

      (iv) Data used (e.g. material laws, boundary conditions, load assumptions) and calculation results should be documented comprehensively.

2.2. THERMAL ANALYSIS

The range of temperatures to be considered for the components of the package for demonstration of compliance is the range of temperatures likely to be encountered in routine conditions of transport for Type IP-1 and Type IP-2 packages. For Type IP-3 packages, the range of temperatures to be considered is specified in para. 639 of the Transport Regulations.

2.3. CONTAINMENT DESIGN ANALYSIS

The analysis of the containment design should demonstrate compliance with the requirements for preventing the loss or dispersal of radioactive material for routine conditions of transport for all industrial packages and with the requirements for normal conditions of transport for Type IP-2 and Type IP-3 packages.

Containment design analysis is not needed if the structural analysis has demonstrated the integrity of the containment boundary, as applicable, for the condition described in para. 645 of the Transport Regulations for reduction of ambient pressure and para. 621 of the Transport Regulations for increase of pressure differential.

2.4. DOSE RATE ANALYSIS

When performing the dose rate analysis, the following points should be considered:

(a) The dose rates for routine conditions of transport for all types of industrial package and a dose rate increase factor for normal conditions of transport for Type IP-2 and Type IP-3 packages should be assessed to demonstrate compliance with the requirements of the Transport Regulations.

(b) The dose rate analysis should be based on assuming the maximum radioactive content of the package or such content for an industrial package as would create the maximum dose rate at the surface of the package and at specific distances from the surface of the package, as defined in the Transport Regulations.

(c) The dose rate analysis should take into account the most recent International Commission on Radiological Protection recommendations on nuclear decay data for dosimetric calculations (see, e.g., Ref. [7]).

(d) The maximum dose rate and a dose rate increase factor for normal conditions of transport, if applicable, in accordance with paras 523.6 and 624.4 of SSG-26 (Rev. 1) [2], should be determined, taking into account potential amplifying phenomena, such as internal movement of the contents (e.g. due to deficiencies of the retention system inside the package in the case of transport of contaminated tools), or — in the case of packages containing liquids — change in the state of the contents, including segregation and precipitation of the radionuclides.

The following should be taken into account when analysing the points listed above:

(1)   Dose rate analysis should be based on the maximum radioactive contents of the package design, which should be defined by various methods and parameters, such as nuclide specific activities and source terms for gamma and neutron emitters.

(2)   The dose rate limits can be shown to be met by calculations or measurements. If calculation methods are used, the calculations of source terms should take into account the interactions, secondary emissions and neutron multiplication factors, when relevant. If dose rate measurements are used, the source used for the measurements should be representative of the radioactive contents specified in the package design.

(3)   All calculation methods used for dose rate analysis should be verified and validated for the specific conditions of the package design they are applied to.

2.5. CRITICALITY SAFETY ANALYSIS

See Appendix V.

Appendix III

TYPE A PACKAGES

III.1. This appendix provides specific recommendations on the information that should be included in Parts 1 and 2 of the PDSR for Type A packages. Table 3 lists each item of the PDSR, with applicable information and guidance. Further recommendations on Type A packages are provided in SSG-26 (Rev. 1) [2].

III.2. In accordance with para. 801 of the Transport Regulations, Type A packages require a PDSR.

III.3. For packages containing fissile nuclides, in addition to the recommendations of this appendix, see also Appendix V.

III.4. For packages containing 0.1 kg or more of uranium hexafluoride, in addition to the recommendations of this appendix, see also Appendix VI.

III.5. For technical analyses, compliance of the package with the Transport Regulations should be demonstrated by the package designer either directly in item 1.5 of the PDSR or, if necessary, in Part 2 of the PDSR.

III.6. For each of the technical analyses in Part 2 of the PDSR, the following considerations apply:

(a) The package design being evaluated should be uniquely identified by precisely indicating a drawing of the packaging (see item 1.3 of the PDSR), including its revision number, and the specification of the contents (see item 1.2 of the PDSR), including its revision number.

(b) The acceptance criteria for the technical analyses and the package design assumptions relating to geometry or performance characteristics should be defined and justified when necessary. The acceptance criteria should be specified by the designer and should be derived from the criteria established by the regulatory body and from other applicable standards developed to meet the regulatory requirements. The design assumptions refer to the design specifications provided in items 1.2 and 1.3 of the PDSR or to other assumptions derived from the design specifications and used in the technical analyses. All mechanical, thermal and shielding characteristics of each component of the package and the acceptance criteria used in technical analyses should be defined.

(c)   The demonstration of compliance of a Type A package design is required to be accomplished in accordance with para. 701 of the Transport Regulations by any of the following methods or by a combination thereof:

(i)   The results of physical testing of prototypes or models of appropriate scale.

(ii)   Reference to previous satisfactory demonstrations of a sufficiently similar nature. Test results of designs similar to the design under consideration are permissible if the similarity can be demonstrated sufficiently by justification and validation.

(iii) Calculation or reasoned argument, when the calculation procedures are generally agreed to be suitable and conservative. Assumptions should be clearly stated and fully justified, including by physical testing if applicable.

Further recommendations are provided in paras 701.1–701.25 of SSG-26 (Rev. 1) [2].

The methods or the standards used in each analysis specified in items 2.1–2.5 of the PDSR should include a description of the analysis technique used, the limitations and accuracy of this technique and a demonstration of the correct application of the technique for the analysis of the package design.

If computer codes are used for the safety analysis, additional information should be included in the PDSR to demonstrate that the code is verified and validated for the specific field of use. The justification for the applicability of these codes should include a statement of possible sources of error and/or uncertainty relative to the effects of the operating platform (computer) used and of modelling assumptions and simplifications, as well as of any other parameter influencing the calculated results and a sensitivity analysis.

(d)   The performance characteristics of the package design should be assessed. More than one sequence of tests might need to be considered to ensure that the various safety functions to be fulfilled by different components of the package design comply with the regulatory requirements. Other hazards (e.g. corrosion, combustion, pyrophoricity or other chemical reactions, radiolysis, phase changes) should be analysed, as necessary, if they have a consequential effect on the safety functions of the package.

(e)   The results of the technical analyses should be compared with the acceptance criteria, and package design assumptions and regulatory compliance should be justified accordingly.

III.7. The following items listed in Fig. 1 and para. 2.3 of this Safety Guide are not relevant to and not needed for Type A packages and, therefore, are not included in Table 3:

— Part 1:
  • Conditions for technical analyses.
— Part 2:
  • Other analyses.

APPENDIX III
TABLE 3. PACKAGE DESIGN SAFETY REPORT FOR TYPE A PACKAGES

Table of contents of the package design safety report (PDSR)

The contents of the PDSR, Part 1 and Part 2, should be listed here, including the revision number of each individual document included in the PDSR

Part 1

1.1. ADMINISTRATIVE INFORMATION

The following administrative information should be provided:

(a) Colloquial name of the package, if applicable;
(b) Identification of the package designer (e.g. name, address, contact details);
(c) Type of package;
(d) Packaging and/or package design identification and restrictions on packaging serial number(s), if applicable;
(e) Modes of transport for which the package is designed, and any operational restrictions associated with any mode of transport;
(f) Reference to the applicable regulations for the specific package design, including the edition of the Transport Regulations to which the package design is referring.

1.2. SPECIFICATION OF THE CONTENTS

A detailed description of the permitted contents of the package design should be provided, stating — at a minimum — the following information, as applicable:

(a)   The general nature of contents (e.g. irradiated fuel, metallurgical specimens, radiographic source).

(b)   The nuclides and/or nuclide composition, including progeny radionuclides.

(c)   The A_1 and A_2 values of the radionuclides to be carried in the package. The A_1 and A_2 values for radionuclides that are not listed in table 2 of the Transport Regulations are required to be determined in accordance with paras 403–407 of the Transport Regulations and may be subject to multilateral approval in accordance with para. 403 of the Transport Regulations.

(d)   The physical and chemical state (additional design specifications are applicable to liquid and gas contents), geometric shape, arrangement, irradiation parameters and material specifications.

(e)   The type and characteristics of the radiation emitted by the contents of the package.

(f)   Limitations on activity, mass and activity concentrations, and heterogeneities in the distribution of the nuclides. Compliance is required with the activity limits for Type A packages in accordance with paras 429 and 430 of the Transport Regulations.

(g)   A valid certificate for special form radioactive material or low dispersible radioactive material, when such material is included in the package.

(h)   The mass of fissile material, the nuclides and the enrichment of the contents (see also Appendix V).

(i)   Other dangerous properties of the contents. In accordance with para. 618 of the Transport Regulations, any other dangerous properties (subsidiary hazards) of the contents of the package are required to be taken into account in the package design to be in compliance with the relevant transport regulations for dangerous goods. Additional information on design requirements for dangerous goods in accordance with the subsidiary hazard can be found in chapter 3.3, Special Provision 172, of Ref. [3].

(j)   Other limitations on the contents (e.g. moisture, presence of acid). Safety relevant limits for non-radioactive materials (e.g. materials subject to radiolysis) — for example, limits in terms of material composition, density, form or location within the package, or restrictions on relative quantities of materials — should be stated.

1.3. SPECIFICATION OF THE PACKAGING

The packaging design should be defined to the extent necessary to demonstrate compliance with the Transport Regulations; the following information should be included, as applicable:

(a)   Design drawings.

(b)   The overall dimensions, the maximum mass of the package when fully loaded, and the mass of the empty packaging.

(c)   A list of all packaging components important to safety and their materials, including the material specifications of the packaging components.

(d)   The maximum normal operating pressure (particularly in the case of air transport).

(e)   A description of the packaging body, lid (e.g. closure mechanism and tamper indicating features), internal arrangements, and components for lifting and tie-down.

(f)   A description of the protection against corrosion.

(g)   A description of the protection against contamination.

(h)   A description of the packaging components of the containment system, including the definition of the containment boundary and the special features for liquid (see para. 650 of the Transport Regulations). Special form radioactive material may be a component of the containment system in accordance with para. 642 of the Transport Regulations, if applicable (see also item 1.2(g) of the PDSR).

(i)   A description of the packaging components required for shielding.

(j)   A description of the shock absorbing components.

(k)   Testing specifications and controls before first use to transport radioactive material. This ensures compliance of the fabrication with the design and allows acceptance of the specimen before its first use. See also para. 501 of the Transport Regulations.

For radioactive material that has other dangerous properties, see para. 4.1.9.1.5 of Ref. [3]. The packing instructions as specified in Ref. [3] (or in the applicable regulations for the carriage of dangerous goods) appropriate to the other dangerous properties of the radioactive material should be complied with, with the exception of packages containing 0.1 kg or more of uranium hexafluoride.

1.4. AGEING CONSIDERATIONS

Depending on the package design, the information relating to ageing considerations can also be provided by the package designer directly in the table mentioned in item 1.5.

For packaging used once for a single transport and not intended for shipment after storage, this item of the PDSR should be left blank.

For all other packaging, this item of the PDSR should include the following information:

(a) The intended conditions of use of the package that might influence ageing;
(b) The potential ageing mechanisms relevant to the package design, taking into account the intended conditions of use of the package;
(c) Operational measures (including maintenance and inspection activities before shipment) to monitor and limit the ageing effects;
(d) Analysis of the influence of ageing of the packaging and contents on the design assumptions for demonstration of compliance with the regulations, including the technical analyses in Part 2 of the PDSR, considering the specified intended use conditions, ageing mechanisms and operational measures.

Further recommendations are provided in paras 613A.1–613A.5 of SSG-26 (Rev. 1) [2].

1.5. COMPLIANCE WITH REGULATORY REQUIREMENTS

The PDSR should include a complete list of all paragraphs of the Transport Regulations and other international or national regulations applicable to the package design.

Compliance of the package design with these regulations should be demonstrated in the PDSR. This could be done using a table (or any other written format) linking the appropriate items of the PDSR, where compliance is demonstrated, to the applicable paragraphs of the regulations.

The applicable paragraphs of the Transport Regulations for Type A packages are provided in a matrix in Annex I.

1.6. PACKAGE OPERATIONS

The minimum specifications for the following activities should be fully defined, as applicable:

(a) Assembly of the packaging components, including demonstration of compliance with para. 637 of the Transport Regulations.
(b) Loading and unloading of the package contents.
(c) Testing and controls before each shipment. The methods used for operational controls and tests, in particular those required in accordance with paras

502, 503(a), 508, 523 and 526–528 of the Transport Regulations, should be detailed.

(d) Handling and tie-down. Specifications on bolt torquing and number of transport cycles (to be used in fatigue analysis) for each mode of transport should be included, if applicable.

(e) Any proposed supplementary equipment and operational controls to be applied during transport.

In addition to the radioactive properties, any other dangerous properties of the contents of the package are required to be taken into account (see para. 507 of the Transport Regulations).

If written procedures with a detailed description of these activities are available, then reference should be made to these procedures.

1.7. MAINTENANCE

The minimum specifications for the following activities should be fully defined, as applicable:

(a) Maintenance and inspection before each shipment;
(b) Maintenance and inspection at periodic intervals throughout the lifetime use of the packaging and/or package.

When developing the maintenance specifications to be included in this item of the PDSR, the following points should be taken into account:

(1) Ageing mechanisms during storage should be considered, when applicable.
(2) For single use packages, periodic maintenance does not need to be considered.
(3) If written procedures with details of the package maintenance are available, then reference should be made to these procedures.

1.8. GAP ANALYSIS PROGRAMME

For packages that are to be used for shipment after storage, the PDSR should include a gap analysis programme describing a systematic procedure for periodic evaluation of changes of regulations, changes in technical knowledge and changes in the state of the package design during storage (see also paras 613A.5, 809.3 and 809.4 of SSG-26 (Rev. 1) [2] and Ref. [8]).

1.9. MANAGEMENT SYSTEM

The PDSR should specify the management system established and implemented by the package designer, in accordance with para. 306 of the Transport Regulations, to demonstrate compliance with the relevant provisions of the Transport Regulations.

The management system should be commensurate with the complexity of the package design and should include a reliable document control system.

Further recommendations on the management system for transport are provided in TS-G-1.4 [6].

1.10. PACKAGE ILLUSTRATION

A reproducible illustration should be provided showing the make-up of the package, including shock absorbers and internal arrangements, if applicable.

The illustration should indicate at least the overall outside dimensions and the mass of the package when empty and when loaded.

Part 2

2.1. STRUCTURAL ANALYSIS

The assessment of the mechanical behaviour (including, as applicable, analysis of fatigue, brittle fracture and creep) for routine and normal conditions of transport should include the following:

(a)   The components of the containment system. This may include special form radioactive material, if applicable, as established in para. 642 of the Transport Regulations.
(b)   The package components that provide radiation shielding.
(c)   Any other package components (e.g. shock absorbing components) whose performance may have a consequential effect on (a) or (b).
(d)   The lifting attachments (see paras 608 and 609 of the Transport Regulations).
(e)   The packaging attachments used to restrain the package within its conveyance for routine conditions of transport.

If the package is to be transported by air (see paras 619–621 of the Transport Regulations), the structural analysis of the containment system should take into account ambient temperatures and pressures likely to be encountered in routine conditions of transport as well as the specific temperature and pressure requirements for air transport. Further recommendations are provided in paras 621.2 and 621.3 of SSG-26 (Rev. 1) [2].

Attention should be paid to ensuring that nuts, bolts and other retention devices maintain their safety functions during routine conditions of transport, even after repeated use. Further recommendations are provided in para. 613.1 of SSG-26 (Rev. 1) [2].

When performing the structural analysis for Type A packages, the following points should be considered:

(1) General considerations:
   (i)   The mechanical properties of the materials considered in the safety demonstration should be representative of the range of mechanical properties of the package components, with account taken of the temperatures likely to be encountered during routine conditions of transport (see para. 639 of the Transport Regulations).
   (ii) The impacts on the package behaviour due to variations in the shock absorbing properties of the shock absorber material (e.g. wood, polymers, plaster, concrete) with temperatures likely to be encountered during routine conditions of transport should be analysed.
   (iii) The safety against brittle fracture at temperatures likely to be encountered during routine conditions of transport of components of the containment system made of potentially brittle materials (e.g. ferritic steels, cast iron) should be analysed, if necessary.
   (iv) The strength of lid bolts should be verified for all drop orientations.
   (v)   The internal components (e.g. content, basket, cage) should be assessed to verify that they are not liable to damage the containment system.
   (vi) The condition of the containment system should be assessed to demonstrate compliance with the specifications in item 2.3 of the PDSR for the temperature range likely to be encountered during routine conditions of transport.
   (vii) Phenomena such as radiolysis, internal pressure elevation, internal inflammation or explosion, physical changes and chemical reactions should be considered when analysing the maximum pressure.

(2) Considerations for experimental mechanical testing:

    (i) The package orientations should be determined in accordance with para. 722.4 of SSG-26 (Rev. 1) [2]. The orientations should maximize the loading of the package (in terms of stress, strain, acceleration and deformation), with consideration of the different package components (e.g. cask body, lid system, shock absorber) and of the protection objectives (i.e. containment and shielding).

    (ii) For reduced scale models, geometry and material properties similar to the original design, or conservative geometry and material properties, should be used.

    (iii) The results of the drop test with reduced scale models should be assessed to guarantee that they cover, or are transferable to, the original design.

    (iv) The representativeness of drop tests performed with reduced scale models should be demonstrated.

    (v) The experimental mechanical tests should be conducted and reported in accordance with the management system. The test report should address the verification of the package before testing, the description of the test site, the equipment used for measurements and its calibration data, and the results of the measurements performed. This report should also contain pictures showing and explaining the conditions under which the tests were performed and their results.

(3) Considerations for calculations:

    (i) See point (2)(i) above.

    (ii) Validated computer codes should be used. Input parameters (e.g. material laws, characteristic values, boundary conditions) should describe sufficiently and precisely the real technical and/or physical problems, and the use of these parameters should be justified.

    (iii) If uncertainties exist regarding important input parameters (e.g. material laws), conservative design calculations — including the possible range of material properties — should be performed.

    (iv) Data used (e.g. material laws, boundary conditions, load assumptions) and calculation results should be documented comprehensively.

2.2. THERMAL ANALYSIS

The range of temperatures to be considered for the components of the package for demonstration of compliance is the range of temperatures specified in para. 639 of the Transport Regulations.

2.3. CONTAINMENT DESIGN ANALYSIS

The analysis of the containment design should demonstrate compliance with the requirements for preventing the loss or dispersal of radioactive material for routine and normal conditions of transport for all Type A packages. For Type A packages containing liquids or gases, additional drop tests are required in accordance with paras 650 and 651 of the Transport Regulations.

Attention should be paid to defining precisely the contents of the package, because assumptions and demonstrations for the containment design analysis can vary depending on the contents.

Where special form radioactive material constitutes part of the containment system, consideration should be given to the appropriate performance of the special form radioactive material for routine and normal conditions of transport.

Containment analysis is not needed if the structural analysis has demonstrated the integrity of the containment boundary, as applicable, for the condition described in para. 645 of the Transport Regulations for reduction of ambient pressure and para. 621 of the Transport Regulations for increase of pressure differential.

2.4. DOSE RATE ANALYSIS

When performing the dose rate analysis, the following points should be considered:

(a) The dose rates for routine conditions of transport and a dose rate increase factor for normal conditions of transport should be assessed to demonstrate compliance with the requirements of the Transport Regulations.

(b) The dose rate analysis should be based on assuming the maximum radioactive content of the package or such content for a Type A package as would create the maximum dose rate at the surface of the package and at specific distances from the surface of the package, as defined in the Transport Regulations.

(c) The dose rate analysis should take into account the most recent International Commission on Radiological Protection recommendations on nuclear decay data for dosimetric calculations (see, e.g., Ref. [7]).

(d) The maximum dose rate and a dose rate increase factor for normal conditions of transport, if applicable in accordance with paras 523.6 and 624.4 of SSG-26 (Rev. 1) [2], should be determined, taking into account potential amplifying phenomena, such as internal movement of the contents (e.g. due to deficiencies of the retention system inside the package in the case of transport of contaminated tools), or — in the case of packages containing liquids — change in the state of the contents, including segregation and precipitation of the radionuclides.

The following should be taken into account when analysing the points listed above:

(1) Dose rate analysis should be based on the maximum radioactive contents of the package design, which should be defined by various methods and parameters, such as nuclide specific activities and source terms for gamma and neutron emitters.

(2) The dose rate limits can be shown to be met by calculations or measurements. If calculation methods are used, the calculations of the source terms should take into account the interactions, secondary emissions and neutron multiplication factors, when relevant. If dose rate measurements are used, the source used for the measurements should be representative of the radioactive contents specified in the package design.

(3) Dose rate analysis should be performed in such a way that the areas of the package surface with the maximum dose rates are identified and analysed. These areas include trunnion areas, areas containing gaps that allow the radiation to pass without being attenuated, and other areas with the potential for increased dose rates due to the design of the package.

(4) All calculation methods used for dose rate analysis should be verified and validated for the specific conditions of the package design they are applied to.

(5) The expected areas for peak dose rates should be specified and checked before shipment.

(6) Proof that the sources will remain secure for drop test sequence conditions in their storage positions (e.g. in irradiators) should be provided in the structural analysis, if applicable.

2.5. CRITICALITY SAFETY ANALYSIS

See Appendix V.

Appendix IV

TYPE B(U), TYPE B(M) AND TYPE C PACKAGES

IV.1. This appendix provides specific recommendations on the information that should be included in Parts 1 and 2 of the PDSR for Type B(U), Type B(M) and Type C packages. Table 4 lists each item of the PDSR, with applicable information and guidance. Further recommendations on Type B(U), Type B(M) and Type C packages are provided in SSG-26 (Rev. 1) [2].

IV.2. For packages containing fissile nuclides, in addition to the recommendations in this appendix, see also Appendix V.

IV.3. For packages containing 0.1 kg or more of uranium hexafluoride, in addition to the recommendations in this appendix, see also Appendix VI.

IV.4. For technical analyses, compliance of the package with the Transport Regulations should be demonstrated by the package designer either directly in item 1.6 of the PDSR or, if necessary, in Part 2 of the PDSR.

IV.5. For each of the technical analyses in Part 2 of the PDSR, the following considerations apply:

(a)   The package design being evaluated should be uniquely identified by precisely indicating a drawing of the packaging (see item 1.3 of the PDSR), including its revision number, and the specification of the contents (see item 1.2 of the PDSR), including its revision number.

(b)   The acceptance criteria for the technical analyses and the package design assumptions relating to geometry or performance characteristics should be defined and justified when necessary. The acceptance criteria should be specified by the designer and should be derived from the criteria established by the regulatory body and from other applicable standards developed to meet the regulatory requirements. The design assumptions refer to the design specifications provided in items 1.2 and 1.3 of the PDSR or to other assumptions derived from the design specifications and used in the technical analyses. All mechanical, thermal and shielding characteristics of each component of the package and the acceptance criteria used in technical analyses should be defined. The design assumptions should take into account ageing mechanisms, as necessary. Further recommendations are provided in paras 613A.1–613A.4 of SSG-26 (Rev. 1) [2].

(c) The demonstration of compliance of a Type B(U), Type B(M) or Type C package design is required to be accomplished in accordance with para. 701 of the Transport Regulations by any of the following methods or by a combination thereof:

(i) The results of physical testing of prototypes or models of appropriate scale. When a programme for physical testing of prototypes or models of appropriate scale is implemented for a specific package design to be approved by the competent authority, the competent authority should be notified of the programme before the testing and should be allowed to witness the testing.

(ii) Reference to previous satisfactory demonstrations of a sufficiently similar nature. Test results of designs similar to the design under consideration are permissible if the similarity can be demonstrated sufficiently by justification and validation.

(iii) Calculation or reasoned argument, when the calculation procedures are generally agreed to be suitable and conservative. Assumptions should be clearly stated and fully justified, including by physical testing if applicable.

Further recommendations are provided in paras 701.1–701.25 of SSG-26 (Rev. 1) [2].

The methods or the standards used in each analysis specified in items 2.1–2.6 of the PDSR should include a description of the analysis technique used, the limitations and accuracy of this technique and a demonstration of the correct application of the technique for the analysis of the package design.

If computer codes are used for the safety analysis, additional information should be included in the PDSR to demonstrate that the code is verified and validated for the specific field of use. The justification for the applicability of these codes should include a statement of possible sources of error and/or uncertainty relative to the effects of the operating platform (computer) used and of modelling assumptions and simplifications, as well as of any other parameter influencing the calculated results and a sensitivity analysis.

(d) The performance characteristics of the package design should be assessed. More than one sequence of tests might need to be considered to ensure that the various safety functions to be fulfilled by different components of the package design comply with the regulatory requirements. Other hazards (e.g. corrosion, combustion, pyrophoricity or other chemical reactions, radiolysis, phase changes) should be analysed, as necessary, if they have a consequential effect on the safety functions of the package.

(e)   The results of the technical analyses should be compared with the acceptance criteria, and package design assumptions and regulatory compliance should be justified accordingly.

APPENDIX IV
TABLE 4. PACKAGE DESIGN SAFETY REPORT FOR TYPE B(U),
TYPE B(M) AND TYPE C PACKAGES

Table of contents of the package design safety report (PDSR)

The contents of the PDSR, Part 1 and Part 2, should be listed here, including the revision number of each individual document included in the PDSR.

Part 1

1.1. ADMINISTRATIVE INFORMATION

The following administrative information should be provided:

(a)   Colloquial name of the package, if applicable;
(b)   Identification of the package designer (e.g. name, address, contact details);
(c)   Type of package;
(d)   UN number;
(e)   Packaging and/or package design identification and restrictions on packaging serial number(s), if applicable;
(f)   Modes of transport for which the package is designed, and any operational restrictions associated with any mode of transport;
(g)   Reference to the applicable regulations for the specific package design, including the edition of the Transport Regulations to which the package design is referring.

1.2. SPECIFICATION OF THE CONTENTS

The description of the contents and of their physical and chemical forms and radionuclides should be sufficiently precise to allow demonstration of compliance with the requirements for containment of the radioactive contents, control of the external dose rate and protection of damage caused by heat.

The description should include all dimensions (drawings), material characteristics and mechanical properties used to demonstrate the required safety performance.

The properties of materials should be given for temperatures ranging from –40°C (or another temperature for Type B(M) packages in accordance with para. 667 of the Transport Regulations) to the maximum temperature in normal conditions of transport. The properties of materials of components expected to maintain their safety function under the thermal test should be given for a range of temperatures reachable during such a test.

The description may include the total numbers of A_1 or A_2 in the contents.

There are additional design requirements depending on the activity (see, e.g., para. 660 of the Transport Regulations).

Compliance with the activity limits for Type B(U) and Type B(M) packages, if transported by air, in accordance with para. 433 of the Transport Regulations, should be considered.

A detailed description of the permitted contents of the package design should be provided, stating — at a minimum — the following information, as applicable:

(a) The general nature of contents (e.g. irradiated fuel, metallurgical specimens, radiographic source).
(b) The nuclides and/or nuclide composition, including progeny radionuclides.
(c) The A_1 and A_2 values of the radionuclides to be carried in the package. The A_1 and A_2 values for radionuclides that are not listed in table 2 of the Transport Regulations are required to be determined in accordance with paras 403–407 of the Transport Regulations and may be subject to multilateral approval in accordance with para. 403 of the Transport Regulations.
(d) The physical and chemical state, geometric shape, arrangement, irradiation parameters (if applicable, the maximum burnup and minimum cooling time), moisture content and material specifications.
(e) The type and characteristics of the radiation emitted by the contents of the package.
(f) Limitations on activity, mass and concentrations, and heterogeneities in the distribution of the nuclides.
(g) A valid certificate for special form radioactive material or low dispersible radioactive material, when such material is included in the package.
(h) Limitations in the heat generation rate of the contents.

(i)   The mass of fissile material, the nuclides and the enrichment of the contents (see also Appendix V).

(j)   Other dangerous properties of the contents. In accordance with para. 618 of the Transport Regulations, any other dangerous properties (subsidiary hazards) of the contents of the package are required to be taken into account in the package design to be in compliance with the relevant transport regulations for dangerous goods. Additional information on the design requirements for dangerous goods in accordance with the subsidiary hazard can be found in chapter 3.3, Special Provision 172, of Ref. [3].

(k)   Other limitations on the contents (e.g. moisture, presence of acid). Safety relevant limits for non-radioactive materials (e.g. materials subject to radiolysis) — for example, limits in terms of material composition, density, form or location within the package, or restrictions on relative quantities of materials — should be stated.

1.3. SPECIFICATION OF THE PACKAGING

The packaging design should be defined to the extent necessary to demonstrate compliance with the Transport Regulations; the following information should be included, as applicable:

(a)   Design drawings.

(b)   The overall dimensions, the maximum mass of the package when fully loaded, and the mass of the empty packaging (additional configurations may be included, depending on the operating conditions).

(c)   A list of packaging components important to safety and their materials, including the specifications and methods of manufacture of the components, specifications for material procurement, welding, other special processes, and non-destructive examination and testing. The properties of materials of components expected to maintain their safety function under the thermal test should be given for a range of temperature reachable during such a test.

(d)   The maximum normal operating pressure.

(e)   A description of the packaging body, lid (e.g. closure mechanism and tamper indicating features), internal arrangements, and components for lifting and tie-down.

(f)   A description of the protection against corrosion.

(g)   A description of the protection against contamination.

(h)   A description of the packaging components for heat dissipation.

(i)   A description of the packaging components of the containment system, including the definition of the containment boundary. Special form radioactive material may be a component of the containment system in accordance with para. 642 of the Transport Regulations, if applicable (see also item 1.2(g) of the PDSR).

(j)   A description of the packaging components required for shielding.

(k)   A description of the shock absorbing components.

(l)   A description of the packaging components for thermal protection.

(m)   Testing specifications and controls before first use to transport radioactive material. This ensures compliance of the fabrication with the design and allows acceptance of the specimen before its first use. See also para. 501 of the Transport Regulations.

1.4. AGEING CONSIDERATIONS

Depending on the package design, the information relating to ageing considerations can also be provided by the package designer directly in the table mentioned in item 1.6.

For packaging used once for a single transport and not intended for shipment after storage, this item of the PDSR should be left blank.

For all other packaging, this item of the PDSR should include the following information:

(a)   The intended conditions of use of the package that might influence ageing;

(b)   The potential ageing mechanisms relevant to the package design, taking into account the intended conditions of use of the package;

(c)   Operational measures (including maintenance and inspection activities before shipment) to monitor and limit the ageing effects;

(d)   Analysis of the influence of ageing of the packaging and contents on the design assumptions for demonstration of compliance with the regulations, including the technical analyses in Part 2 of the PDSR, considering the specified intended use conditions, ageing mechanisms and operational measures.

Further recommendations are provided in paras 613A.1–613A.5 of SSG-26 (Rev. 1) [2], and additional information on ageing considerations can be found in Ref. [8].

1.5. CONDITIONS FOR TECHNICAL ANALYSES

This item of the PDSR should describe the main design principles and performance characteristics of the package design to meet the safety requirements (e.g. containment, heat removal, dose rates) of the Transport Regulations.

This item should summarize the analyses performed in Part 2 of the PDSR and describe how analysis assumptions and data used for the safety analysis — especially regarding release of radioactive material and dose rates — are derived from the design and the behaviour of the package for routine, normal and accident conditions of transport, also taking into account ageing mechanisms (see item 1.4 of the PDSR).

This item will help ensure that the package design and the various parts of the safety demonstration are compatible with one another.

1.6. COMPLIANCE WITH REGULATORY REQUIREMENTS

The PDSR should include a complete list of all paragraphs of the Transport Regulations and other international or national regulations applicable to the package design.

Compliance of the package design with these regulations should be demonstrated in the PDSR. This could be done using a table (or any other written format) linking the appropriate items of the PDSR, where compliance is demonstrated, to the applicable paragraphs of the regulations.

The applicable paragraphs of the Transport Regulations for Type B(U), Type B(M) and Type C packages are provided in a matrix in Annex I.

1.7. PACKAGE OPERATIONS

The minimum specifications for the following activities should be fully defined, as applicable:

(a)   Assembly of the packaging components, including demonstration of compliance with para. 637 of the Transport Regulations.
(b)   Loading and unloading of the package contents.
(c)   Testing and controls before each shipment:
    (i)   The methods used for operational controls and tests, in particular those required in accordance with paras 502, 503, 508, 523 and 526–528 of the Transport Regulations, should be detailed.
    (ii)   The measures aimed at preventing the presence of unauthorized objects (e.g. tools, small pieces of plastic, worn gaskets) in the package should be defined.
    (iii)   The control of all void spaces of the package (i.e. cavity and other spaces), in particular regarding water penetration, should be specified.
    (iv)   For drying operations, the method used should prevent formation of ice.

(v) For leaktightness testing, qualified methods should be implemented (see item 2.3 of the PDSR). For packages that are or have been in contact with water, it should be demonstrated that the presence of water does not impair the validity of the leaktightness testing by sealing the leakage paths.

(vi) The absence of defects should be ensured by a specific inspection procedure that has been appropriately qualified.

(vii) The control of tightening torques of the bolts and of the correct position of the lid, and the adjustment of the internal atmosphere and pressure, should be specified.

(d) Handling and tie-down. Specifications on bolt torquing and number of transport cycles (to be used in fatigue analysis) for each mode of transport should be included, if applicable.

(e) Estimation of the correction factor to be applied to the dose rate and the transport index to take into account any amplifying phenomena (paras 523.6 and 624.4 of SSG-26 (Rev. 1) [2]).

(f) Any proposed supplementary equipment and operational controls to be applied during transport and, if applicable, during storage before transport, including those that might influence ageing mechanisms.

In addition to the radioactive properties, any other dangerous properties of the contents of the package are required to be taken into account (see para. 507 of the Transport Regulations).

If written procedures with a detailed description of these activities are available, then reference should be made to these procedures.

1.8. MAINTENANCE

The minimum specifications for the following activities should be fully defined, as applicable:

(a) Maintenance and inspection before each shipment;

(b) Maintenance and inspection at periodic intervals throughout the lifetime use of the packaging and/or package.

Periodic maintenance and inspection activities should be detailed and may include the following activities and tests, depending on the package design:

(1) Visual inspections and measurements (including tie-down and handling attachments);

(2)   The control of all void spaces of the package (i.e. cavity and other spaces), in particular regarding water penetration;

(3)   Weld examinations;

(4)   Structural (including tie-down and handling attachments) and pressure tests;

(5)   Leakage tests;

(6)   Component and material tests (e.g. screws, bolts, welds, gaskets, seals, wood, foam, resin);

(7)   Shielding tests;

(8)   Thermal verification tests.

The specified periodicity of replacement of the packaging components should take into account any reduction in efficiency due to wear, corrosion, ageing or change in seal compression with time.

The justification of the periodicity of activities, when needed, should be included in this item of the PDSR.

If written procedures with details of the maintenance activities are available, then reference should be made to these procedures.

1.9. GAP ANALYSIS PROGRAMME

For packages that are to be used for shipment after storage, the PDSR should include a gap analysis programme describing a systematic procedure for periodic evaluation of changes of regulations, changes in technical knowledge and changes in the state of the package design during storage (see also paras 613A.5, 809.3 and 809.4 of SSG-26 (Rev. 1) [2] and Ref. [8]).

1.10. MANAGEMENT SYSTEM

The PDSR should specify the management system established and implemented by the package designer, in accordance with para. 306 of the Transport Regulations, to demonstrate compliance with the relevant provisions of the Transport Regulations.

The management system should cover the following activities:

(a)   Package design, PDSR, documentation, records and use of computer codes;

(b)   Manufacture and testing of the packaging;

(c)   Operation (i.e. preparation, loading, carriage, storage in transit, shipment after storage, unloading and receipt);

(d)   Maintenance, repair and inspection of the packaging.

The management system should be commensurate with the complexity of the package design and should include a reliable document control system.

The management system should include descriptions of the actions to be performed to check compliance of the documents relating to package operation (e.g. manufacturing, operation or maintenance manuals) with the PDSR and should also cover the management of deviations detected in the framework of any transport activity.

For all components important to safety, the PDSR should define the parameters to be guaranteed to ensure compliance with the package design, and therefore safety, and the level of controls required during manufacturing and maintenance.

Further recommendations on the management system for transport are provided in TS-G-1.4 [6].

1.11. PACKAGE ILLUSTRATION

A reproducible illustration should be provided showing the make-up of the package, including shock absorbers, devices for thermal protection, and internal arrangements, if applicable.

The illustration should indicate at least the overall outside dimensions and the mass of the package when empty and when loaded.

Part 2

2.1. STRUCTURAL ANALYSIS

The assessment of the mechanical behaviour (including, as applicable, analysis of thermal stresses, fatigue, brittle fracture and creep) for routine, normal and accident conditions of transport should include the following:

(a) The components of the containment system. This may include special form radioactive material, if applicable, as established in para. 642 of the Transport Regulations.
(b) The package components that provide radiation shielding.
(c) Any other package components (e.g. shock absorbing components, packaging components that provide heat dissipation) whose performance may have a consequential effect on (a) or (b).

(d) The lifting attachments (see paras 608 and 609 of the Transport Regulations).

(e) The packaging attachments used to restrain the package within its conveyance for routine conditions of transport.

If the package is to be transported by air (see paras 619–621 of the Transport Regulations), the structural analysis of the containment system should take into account ambient temperatures and pressures likely to be encountered in routine conditions of transport as well as the specific temperature and pressure requirements for air transport. Further recommendations are provided in paras 621.2 and 621.3 of SSG-26 (Rev. 1) [2].

Attention should be paid to ensuring that nuts, bolts and other retention devices maintain their safety functions during routine conditions of transport, even after repeated use. Further recommendations are provided in para. 613.1 of SSG-26 (Rev. 1) [2].

When performing the structural analysis for Type B(U), Type B(M) and Type C packages, the following points should be considered:

(1) General considerations:

(i) The mechanical properties of the materials considered in the safety demonstration should be representative of the range of mechanical properties of the package components. The analysis should also consider the temperature range of –40°C (or another specified temperature for Type B(M) packages in accordance with para. 667 of the Transport Regulations) to +70°C (see para. 639 of the Transport Regulations) and the temperature range of the package components in normal conditions of transport (see para. 653 of the Transport Regulations).

(ii) The impacts on the package behaviour due to variations in the shock absorbing properties of the shock absorber material (e.g. wood, polymers, plaster, concrete) should be analysed for the temperature range of –40°C (or another specified temperature for Type B(M) packages in accordance with para. 667 of the Transport Regulations) to the maximum temperature in normal conditions of transport, and for the range of moisture conditions likely to be encountered during transport.

(iii) The safety against brittle fracture should be analysed at –40°C (or another specified temperature for Type B(M) packages in accordance with para. 667 of the Transport Regulations) for those components of the containment system made of potentially brittle materials (e.g. ferritic steels, cast iron).

(iv)   The strength of the lid bolts should be verified for all drop orientations.

(v)   Any excursion of stress into the plastic domain should be avoided to the extent possible for containment system components such as bolts and gasket seats, which would need additional complex proofs concerning the mechanics of the rupture or the maintenance of sufficient gasket seating.

(vi)   Any possible damage of metallic seals after drops due to vibrations or sliding of the lid should be evaluated.

(vii)   The internal components (e.g. content, basket, cage) should be assessed to verify that they are not liable to damage the containment system. For the evaluation of the impact of internal components on the packaging lid, the maximum possible gap between these components and the lid before the drop should be considered.

(viii)   The condition of the containment system should be assessed to demonstrate compliance with the specifications in item 2.3 of the PDSR for the temperature range concerned, that is, from $-40°C$ (or another specified temperature for Type B(M) packages, in accordance with para. 667 of the Transport Regulations) to the maximum temperature in accident conditions of transport.

(ix)   Retention of sufficient thermal protection should be demonstrated, after the mechanical tests for accident conditions of transport, to guarantee the containment or other components' safety function during the thermal test.

(x)   The effect of the thermal test on the mechanical behaviour of the package (e.g. thermal stresses and strains, thermo-mechanical interactions between package components, interactions of the package components with the contents) should be considered.

(xi)   If the shielding includes components made from lead, the consolidation height of lead (lead slump) after the 9 m drop test should be determined, taking into account the environmental conditions described in paras 656 and 657 of the Transport Regulations.

(xii)   Phenomena such as radiolysis, internal pressure elevation, internal inflammation or explosion, physical changes and chemical reactions, and accident conditions of transport (including the thermal test) should be considered when analysing the maximum pressure.

(xiii)   An appropriate water immersion test, depending on the content activity of the package, should be considered.

(2) Considerations for experimental mechanical testing:
   (i) The package orientations should be determined in accordance with paras 722.4 and 727.5 of SSG-26 (Rev. 1) [2]. The orientations should maximize the loading of the package (in terms of stress, strain, acceleration and deformation), with consideration of the different package components (e.g. cask body, lid system, shock absorber) and of the protection objectives (e.g. containment, shielding).

       For package orientations, the following tests should be considered:
- Tests that maximize the stresses and acceleration (e.g. flat, slap down). The greater the impact area, the harder the impact, under the assumption of constant stiffness per unit area.
- Tests that maximize the deformation (e.g. on corners, on edges). The smaller the impact area, the greater the crushing.
- Tests that maximize the damage to orifices, notably by a puncture bar. The containment components in the orifices are often thin and more liable than the body of the packaging to be damaged by the bar.
- Tests that maximize the risk of perforation by a puncture bar, possibly oblique. If the impacted package surface is oblique with respect to the puncture bar, the initial impact takes place on an edge of the puncture bar and the risk of perforation is much higher.

   (ii) For reduced scale models, geometry and material properties similar to the original design, or conservative geometry and material properties, should be used.

   (iii) The results of the drop test with reduced scale models should be assessed to guarantee that they cover, or are transferable to, the original design.

   (iv) Drop tests performed with reduced scale models should be demonstrated as representative for the following parameters and components:
- Drop heights: It might be necessary to increase the drop heights during testing to simulate the total potential energy that would have been received by the package at full scale. This should be considered for drop tests where the characteristic deformation of the structure is not negligible in comparison to the drop height.
- Appropriate geometry scaling of all components (e.g. lids, nuts and bolts, grooves for the seals).

— Metallic gaskets: The same design, same material and homothetic transformation with regard to elastic restitution should be selected.

— O-rings: The selection should be based on the similarity of the useful elastic restitution, taking into account the compression set. The change of material properties with the temperature conditions should be considered.

— The scaling of tightening torques for bolts of the reduced scale model should take into account the dispersion of friction conditions, the precision of torques and the technical limitations in an exact geometrical and physical scaling of the containment system components.

— Welding seams should be similar in the scale model and package design.

— For reduced scale model drop testing with significant deformations of shock absorbers, the original package performance should be carefully justified.

(v)   The experimental mechanical tests should be conducted and reported in accordance with the management system. The test report should address the verification of the package before testing, the description of the test site, the equipment used for measurements and its calibration data, and the results of the measurements performed. This report should also contain pictures showing and explaining the conditions under which the tests were performed and their results.

(3)   Considerations for calculations:

(i)   See point (2)(i) above.

(ii)   Validated computer codes should be used. Input parameters (e.g. material laws, characteristic values, boundary conditions) should describe sufficiently and precisely the real technical and/or physical problems, and the use of these parameters should be justified.

(iii)   If uncertainties exist regarding important input parameters (e.g. material laws), conservative design calculations — including the possible range of material properties — should be performed.

(iv)   Data used (e.g. material laws, boundary conditions, load assumptions) and calculation results should be documented comprehensively.

2.2. THERMAL ANALYSIS

The temperature of the accessible surfaces of a package under the conditions defined in para. 654 or 655 of the Transport Regulations should be determined.

The temperature of the package components should be assessed in normal conditions of transport (see para. 653 of the Transport Regulations) and accident conditions of transport (see para. 659(b) of the Transport Regulations). The thermal analysis should include the thermal behaviour of the following components:

(a) The components of the containment system;
(b) The components of radiation shielding;
(c) The package components whose performance will have a consequential effect on (a) or (b).

The following should be considered:

(1) The effects of insolation over a period of 12 hours in accordance with para. 657 of the Transport Regulations.
(2) The presence of protective systems liable to oppose heat dissipation in normal conditions of transport. Protective systems to be considered include, as applicable, tarpaulins, canopies, additional screens and outer packaging (e.g. containers, boxes).
(3) The solar insolation before and after the thermal test, as defined in para. 728 of the Transport Regulations.
(4) The justification for simplifying the assumptions (e.g. absence of trunnions) used for calculation under normal and accident conditions of transport.
(5) The analysis of the package in accident conditions of transport under the position (i.e. horizontal or vertical) least resistant to thermal tests.
(6) The value of the surface absorptivity coefficient of the package. The surface absorptivity coefficient should not be lower than 0.8 (see para. 728(a) of the Transport Regulations) during and after the thermal test to account for deposits on the package surface. The surface absorptivity coefficient should also not be lower than the possible maximum value of the emissivity coefficient in routine conditions of transport.
(7) The evaluation of the minimum and maximum temperatures of the various components of the packaging, taking account of all the possible positions for the radioactive contents.
(8) The profile of heat power based on the burnup distribution in irradiated fuels.
(9) The justification that the temperature measurements were performed at thermal equilibrium, when thermal analysis for the conditions specified in paras 654 and 655 of the Transport Regulations is based on test results.
(10) The justification that the concentration of oxygen present in the furnace is controlled and in conformity with that obtained in a hydrocarbon fuel–air fire, when the thermal test is conducted in a furnace and when it is noted that

some package components burn. In addition, control of heat input should be considered thoroughly.

(11) The influence of combustible materials that generate additional heat input and affect the fire duration beyond the thermal test.

(12) The safety margins on temperature results derived using numerical modelling commensurate with the uncertainty associated with the numerical model.

(13) Demonstration that the spare volume in the gasket grooves allows for gasket thermal expansion in the conditions specified in para. 653 of the Transport Regulations and accident conditions of transport, unless appropriate justification is provided.

2.3. CONTAINMENT DESIGN ANALYSIS

All possible releases, in the form of gases, liquids, solids or aerosols, through leaks or by permeation, should be included in the containment design analysis.

The technical assessment of the containment design should demonstrate compliance with the release criteria for normal and accident conditions of transport, and the following points should be considered:

(a) The mechanical resistance of the irradiated fuel assemblies with respect to the internal pressure should be assessed in item 2.1 of the PDSR. The risk of rupture due to creep of the rods under the effect of internal pressure should be evaluated, taking into account the mechanical properties of the fuel cladding for the temperature conditions in normal conditions of transport and for the burnup of irradiated fuel assemblies in combination with the free drop test.

(b) The analysis of the state of the irradiated fuel assemblies in accident conditions of transport (risk of cracking or rupture of the fuel rod at their ends) should be included in item 2.1 of the PDSR for safety demonstration, if necessary.

(c) The fraction of fission gas released from the fuel material should be justified.

(d) The presence of debris and of aerosols in the package cavity for irradiated fuels in the case of rupture due to the shearing of the rods should be considered.

(e) The formation of aerosols for contents consisting of materials in powder form should be considered for accident conditions of transport.

(f) The long term behaviour of gasket material should be considered (see item 1.4 of the PDSR).

(g) A reduction of ambient pressure to 60 kPa should be considered for evaluation of activity release.

2.4. DOSE RATE ANALYSIS

When performing the dose rate analysis, the following points should be considered:

(a) The dose rates for routine and accident conditions of transport and a dose rate increase factor for normal conditions of transport should be assessed to demonstrate compliance with the requirements of the Transport Regulations.

(b) The dose rate analysis should be based on assuming the maximum radioactive content of the package or such content for a Type B(U), Type B(M) or Type C package as would create the maximum dose rate at the surface of the package and at specific distances from the surface of the package, as defined in the Transport Regulations.

(c) The dose rate analysis should take into account the most recent International Commission on Radiological Protection recommendations on nuclear decay data for dosimetric calculations (see, e.g., Ref. [7]).

(d) The maximum dose rate and a dose rate increase factor for normal conditions of transport, if applicable in accordance with paras 523.6 and 624.4 of SSG-26 (Rev. 1) [2], should be determined, taking into account potential amplifying phenomena, such as internal movement of the contents, or — in the case of packages containing liquids — change in the state of the contents, including segregation and precipitation of the radionuclides.

The following should be taken into account when analysing the points listed above:

(1) Dose rate analysis should be based on the maximum radioactive contents of the package design, which should be defined by various methods and parameters, such as nuclide specific activities and source terms for gamma and neutron emitters.

(2) The dose rate limits can be shown to be met by calculations or measurements. If calculation methods are used, the calculations of source terms should take into account the interactions, secondary emissions and neutron multiplication factors, when relevant. If dose rate measurements are used, the sources used for the measurements should be representative of the radioactive contents specified in the package design, and appropriate dose rate measuring techniques should be used for gamma and neutron radiation, as applicable.

(3) Dose rate analysis should be performed in such a way that the areas of the package surface with the maximum dose rates are identified and analysed. These areas include trunnion areas, areas containing gaps that allow the radiation to pass without being attenuated, and other areas with the potential for increased dose rates due to the package design.

(4)   All calculation methods used for dose rate analysis should be verified and validated for the specific conditions of the package design they are applied to.

(5)   The expected areas for peak dose rates should be specified and checked before shipment.

(6)   Proof that the sources will remain secure for drop test sequence conditions in their storage positions (e.g. in irradiators) should be provided in the structural analysis, if applicable.

(7)   For the materials providing radiation shielding, local melting or combustion during the thermal test should be considered, as determined by the thermal analysis, and the thermal analysis should take into account the effects of penetration or deformation of components by the bar in the mechanical test.

2.5. CRITICALITY SAFETY ANALYSIS

See Appendix V.

2.6. OTHER ANALYSES

Radiolysis and thermolysis phenomena can affect and can be affected by the structural analysis, the thermal analysis, the containment design analysis and the dose rate analysis.

If the package contains wet contents, the internal pressure built up inside the package should be assessed and considered for regulatory transport conditions.

For the assessment of effects (e.g. internal pressure elevation, internal ignition or explosion) concerning radiolysis and/or thermolysis on the performance characteristics of the package design, the following points should be considered:

(a)   If water or hydrocarbonated materials (e.g. cellulose, polymers, aqueous or organic solutions, absorbed humidity) are present in the package, the absence of the risk of accumulation of combustible gases exceeding the limiting concentration for flammability should be demonstrated.

(b)   When the radiolysis phenomenon limits the maximum safe duration of transport, any specified limit for the duration of transport should integrate margins for incidents and emergency response operations.

(c)   If leaking fuel rods are allowed as contents and absence of water has not been demonstrated, contained water should be taken into account.

If applicable, the risk of chemical or physical reactions for materials that react with water or oxygen (e.g. sodium, uranium hexafluoride, plutonium, metallic uranium) should be assessed. Moreover, potential change of phase (i.e. freezing, melting, boiling), precipitation or segregation should be considered.

Appendix V

**ADDITIONAL INFORMATION FOR PACKAGES
CONTAINING FISSILE NUCLIDES**

V.1. This appendix provides specific additional recommendations on the information to be provided in Parts 1 and 2 of the PDSR for packages containing fissile nuclides, as defined in para. 222 of the Transport Regulations. Table 5 lists each item of the PDSR, with applicable information and guidance.

V.2. The recommendations provided in this appendix apply in addition to those items associated with the package type defined by the radioactive properties of the contents (see Appendices I–IV).

V.3. Further recommendations are provided in SSG-26 (Rev. 1) [2], and further guidance can be found in Ref. [9].

V.4. For packages containing 0.1 kg or more of uranium hexafluoride, see also Appendix VI.

The Transport Regulations include the following four groups of provisions for the transport of radioactive material containing fissile nuclides, also shown in the bottom row of Fig. 2:

— Group 1: Transport where the material is not classified as fissile, in accordance with the definition of fissile material;
— Group 2: Transport with exception from UN "FISSILE" dangerous goods classification and criticality safety index (CSI) control;
— Group 3: Transport with UN "FISSILE" dangerous goods classification and CSI control but excepted from competent authority approval as a package design for fissile material;
— Group 4: Transport in a package for which the design is approved by the competent authority to contain fissile material.

V.5. The same package design may be assigned to different groups for different consignments. This should be reflected in the PDSR. For each group, the complete information should be given in the PDSR, as specified in Table 5.

V.6. For Group 1, no additional information is needed in the PDSR.

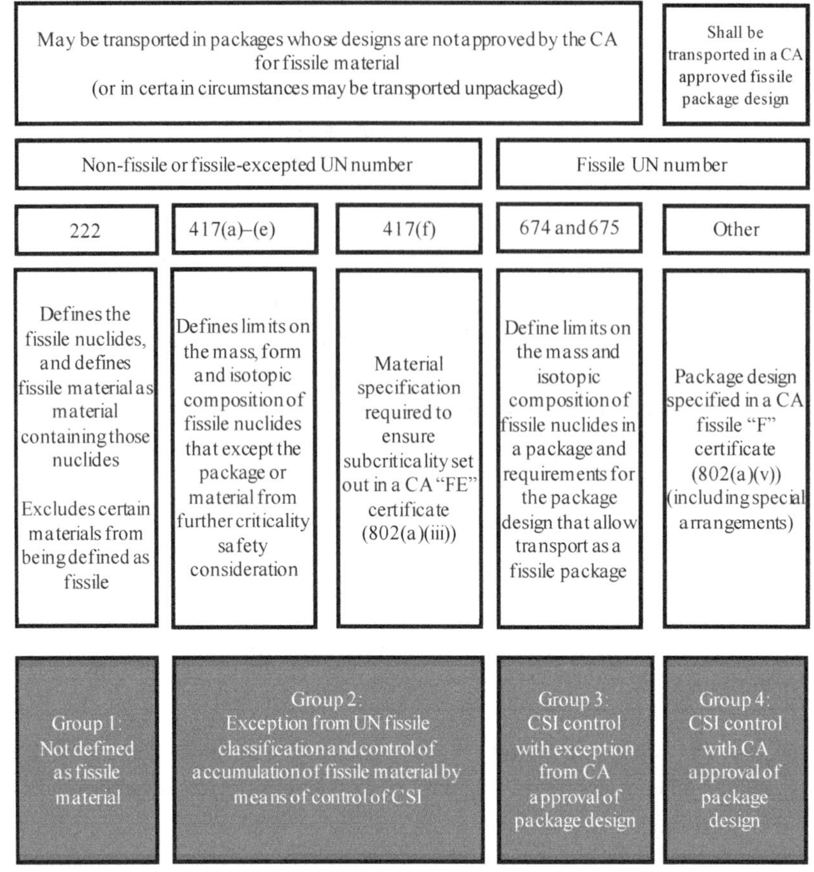

FIG. 2. Overview of the provisions for the transport of fissile material (adapted from Ref. [9], where CA — competent authority; CSI — criticality safety index; FE — fissile-excepted; F — fissile). Paragraph numbers are from the Transport Regulations [1].

V.7. For technical analyses, compliance of the package with the Transport Regulations should be demonstrated by the package designer either directly in item 1.6 of the PDSR or, if necessary, in Part 2 of the PDSR.

V.8. For each of the technical analyses in Part 2 of the PDSR, the following considerations apply:

(a)	The package design being evaluated should be uniquely identified by precisely indicating a drawing of the packaging (see item 1.3 of the PDSR), including its revision number, and the specification of the contents (see item 1.2 of the PDSR), including its revision number.

(b)   The acceptance criteria for the technical analyses and the package design assumptions relating to geometry or performance characteristics should be defined and justified when necessary. The acceptance criteria should be specified by the designer and should be derived from the criteria established by the regulatory body and from other applicable standards developed to meet the regulatory requirements. The design assumptions refer to the design specifications provided in items 1.2 and 1.3 of the PDSR or to other assumptions derived from the design specifications and used in the technical analyses. All mechanical and thermal characteristics of each component of the package and the acceptance criteria used in technical analyses should be defined. The design assumptions should take into account ageing mechanisms, as necessary. Further recommendations are provided in paras 613A.1–613A.4 of SSG-26 (Rev. 1) [2].

(c)   The demonstration of compliance of a package design for fissile material is required to be accomplished in accordance with para. 701 of the Transport Regulations by any of the following methods or by a combination thereof:

(i)   The results of physical testing of prototypes or models of appropriate scale. When a programme for physical testing of prototypes or models of appropriate scale is implemented for a specific package design to be approved by the competent authority, the competent authority should be notified of the programme before the testing and should be allowed to witness the testing. When such a programme of tests is done without competent authority approval but is part of the safety analysis, its validity is to be determined by the competent authority.

(ii)   Reference to previous satisfactory demonstrations of a sufficiently similar nature. Test results of designs similar to the design under consideration are permissible if the similarity can be demonstrated sufficiently by justification and validation.

(iii) Calculation or reasoned argument, when the calculation procedures are generally agreed to be suitable and conservative. Assumptions should be clearly stated and fully justified, including by physical testing if applicable.

Further recommendations are provided in paras 701.1–701.25 of SSG-26 (Rev. 1) [2].

The methods or standards used in each analysis specified in items 2.1–2.4 of the PDSR should include a description of the analysis technique used, the limitations and accuracy of this technique and a demonstration of the correct application of the technique for the analysis of the package design.

If computer codes are used for the safety analysis, additional information should be included in the PDSR to demonstrate that the code is verified and validated for the specific field of use. The justification for the applicability of these codes should include a statement of possible sources of error and/or uncertainty relative to the effects of the operating platform (computer) used and of modelling assumptions and simplifications, as well as of any other parameter influencing the calculated results and a sensitivity analysis.

(d)   The performance characteristics of the package design should be assessed. More than one sequence of tests might need to be considered to ensure that the various safety functions to be fulfilled by different components of the package design comply with the regulatory requirements. Other hazards (e.g. corrosion, combustion, pyrophoricity or other chemical reactions, radiolysis, phase changes) should be analysed, as necessary, if they have a consequential effect on the safety functions of the package.

(e)   The results of the technical analyses should be compared with the acceptance criteria, and package design assumptions and regulatory compliance should be justified accordingly.

V.9.   The following items listed in Fig. 1 and para. 2.3 of this Safety Guide are not relevant to and not needed for packages containing fissile nuclides and, therefore, are not included in Table 5:

— Part 2:
  • Dose rate analysis;
  • Other analyses.

APPENDIX V
TABLE 5. PACKAGE DESIGN SAFETY REPORT FOR PACKAGES
CONTAINING FISSILE NUCLIDES, ADDITIONAL INFORMATION

Part 1

1.1. ADMINISTRATIVE INFORMATION

Group 2: When one of the provisions of para. 417(a)–(f) of the Transport Regulations applies to the package, then a reference to the provision should be added. Especially, when para. 417(f) applies, multilateral approval is required for the exception of the contents.

Group 3: When one of the provisions of para. 674(a)–(c) or of para. 675 applies to the package contents, then a reference to this provision should be added. The colloquial name of the package, if applicable, should be added.

Group 4: The following administrative information should be added, when necessary: type of package.

1.2. SPECIFICATION OF THE CONTENTS

The description of the contents and of their physical and chemical forms and radionuclides should be sufficiently precise to allow demonstration of compliance with the requirements for prevention of criticality.

The following information should be added, as necessary:

Group 2: The mass of fissile nuclides and enrichment, if applicable.

Group 3:

(a) The mass of fissile nuclides and enrichment, if applicable.
(b) Other limitations on the contents, as described in para. 674(d) or 675(c) of the Transport Regulations.
(c) The formula for calculating the CSI in accordance with para. 674 or 675 of the Transport Regulations.

Group 4:

(a) The mass of fissile nuclides and enrichment of the contents, if applicable. A description of quantities of nuclides not defined as fissile but able to sustain chain reaction (e.g. if certain actinides could be present in sufficient quantity or concentration to increase the neutron multiplication factor, their concentrations and/or quantities should be defined).

(b) The CSI and the value of the number N defined in paras 684 and 685 of the Transport Regulations.

(c) Safety relevant limits for non-radioactive materials (e.g. moderators, reflectors), for example limits on material composition, density, form or location within package, or restrictions of relative quantities of materials.

Criticality safety can be very sensitive to the presence and geometrical arrangement of fissile material (e.g. possibility and size of lattice arrangements), moderators (e.g. water, graphite, beryllium, other light elements) and reflectors. This should be taken into account in the description of the permitted and not permitted contents.

1.3. SPECIFICATION OF THE PACKAGING

No additional information is needed for Group 2.

The following information should be added, when necessary:

Group 3: Testing specifications and controls before first use to transport radioactive material. This ensures compliance of the fabrication with the design and allows acceptance of the specimen before its first use. See also para. 501 of the Transport Regulations.

Group 4:

(a) The overall dimensions, the maximum mass of the package when fully loaded, and the mass of the empty packaging (additional configurations may be included, depending on the operating conditions).

(b) A list of packaging components important to safety and their materials, including the specifications and methods of manufacture of the components, specifications for material procurement, welding, other special processes, and non-destructive examination and testing. The properties of materials of components expected to maintain their safety function under the thermal test should be given for a range of temperature reachable during such a test.

(c) A description of the packaging components of the confinement system (e.g. neutron poisons, moderators, flux traps, spacer).

(d) A description of the packaging components for heat dissipation.

(e) A description of the packaging components for thermal protection.

(f) Testing specifications and controls before first use to transport radioactive material. This ensures compliance of the fabrication with the design and allows acceptance of the specimen before its first use. See also para. 501 of the Transport Regulations.

1.4. AGEING CONSIDERATIONS

Depending on the package design, the information relating to ageing considerations can also be provided by the package designer directly in the table mentioned in item 1.6.

For packaging used once for a single transport and not intended for shipment after storage, this item of the PDSR should be left blank.

For all other packaging, this item of the PDSR should include the following information:

(a) The intended conditions of use of the package that might influence ageing;

(b) The potential ageing mechanisms relevant to the package design, taking into account the intended conditions of use of the package;

(c) Operational measures (including maintenance and inspection activities before shipment) to monitor and limit the ageing effects;

(d) Analysis of the influence of ageing of the packaging and contents on the design assumptions for demonstration of compliance with the regulations, including the technical analyses in Part 2 of the PDSR, considering the specified intended use conditions, ageing mechanisms and operational measures.

Further recommendations are provided in paras 613A.1–613A.5 of SSG-26 (Rev. 1) [2].

1.5. CONDITIONS FOR TECHNICAL ANALYSES

No additional information is needed for Groups 2 and 3.

Group 4: This item of the PDSR should describe the main design principles and performance characteristics of the package design to meet the criticality safety requirements of the Transport Regulations.

This item should summarize the analyses performed in Part 2 of the PDSR and describe how analysis assumptions and data used for the criticality safety analysis are derived from the design and the behaviour of the package for routine, normal and accident conditions of transport, also taking into account ageing mechanisms (see item 1.4 of the PDSR).

This item will help ensure that the package design and the various parts of the safety demonstration are compatible with one another.

All assumptions about the state of the package used in the criticality safety assessment for normal and accident conditions of transport should be listed and justified. The state of the components of the confinement system under normal and accident conditions should be derived from the design and the behaviour of the package under test conditions. Otherwise, conservative assumptions should be taken, and their conservatism should be demonstrated.

Often, test conditions leading to the maximum damage in terms of activity release or dose rate increase do not result in the maximum neutron multiplication. Therefore, for the criticality safety assessment, additional tests may have to be considered. For any parameter not justified, the value leading to maximum neutron multiplication should be identified and used in the criticality safety assessment. For cases where complete or partial water filling of cavities is important for criticality safety, the filling states considered and those excluded from the assessment should be described and justified.

1.6. COMPLIANCE WITH REGULATORY REQUIREMENTS

The PDSR should include a complete list of all paragraphs of the Transport Regulations and other international or national regulations applicable to packages containing fissile nuclides.

Compliance of the package design with these regulations should be demonstrated in the PDSR. This could be demonstrated using a table (or any other written format) linking the appropriate items of the PDSR, where compliance is demonstrated, to the applicable paragraphs of the regulations.

The applicable paragraphs of the Transport Regulations for packages containing fissile nuclides are provided in a matrix in Annex I.

1.7. PACKAGE OPERATIONS

The following information should be added, when necessary:

Group 2: Any operational controls to be applied during transport, including consignment and conveyance limits.

Group 3: Specifications for assembling the packaging components, including compliance with the requirements established in para. 637 of the Transport Regulations.

Group 4:

(a) Testing specifications and controls before each shipment:
   (i) The methods used for operational controls and tests, in particular those required in accordance with paras 502 and 503 of the Transport Regulations, should be detailed.
   (ii) For drying operations, the method used should prevent formation of ice.
   (iii) For leaktightness testing, qualified methods should be implemented (see item 2.3 of the PDSR). For packages that are or have been in contact with water, it should be demonstrated that the presence of water does not impair the validity of the leaktightness testing by sealing the leakage paths.
   (iv) The check for the presence of absorber rods or for the selection of inner equipment with the correct neutron absorber content should be specified, if applicable.
   (v) The control of tightening torques of the bolts and of the correct position of the lid should be specified.
(b) Specifications for assembling the packaging components, including compliance with the requirements established in para. 637 of the Transport Regulations.

If written procedures with a detailed description of these activities are available, then reference should be made to these procedures.

1.8. MAINTENANCE

No additional information is needed for Groups 2 and 3.

Group 4: The following information should be added, if necessary:

(a) Periodic maintenance and inspection activities should be detailed and may include the following activities and tests, depending on the package design:
    (i) Visual inspections and measurements (including tie-down and handling attachments);
    (ii) The control of all void spaces of the package (i.e. cavity and other spaces), in particular regarding water penetration;
    (iii) Weld examinations;
    (iv) Structural (including internal enclosures, and tie-down and handling attachments) and pressure tests;
    (v) Component and material tests (e.g. screws, bolts, welds, neutron absorbers, basket).

(b) The specified periodicity of replacement of the packaging components should take into account any reduction in efficiency due to wear, corrosion or ageing.

(c) The justification of the periodicity of activities, when needed, should also be included in this item of the PDSR.

1.9. GAP ANALYSIS PROGRAMME

No additional information is needed for Groups 2 and 3.

Group 4: Additional information about the systematic procedure for periodic evaluation of changes in technical knowledge should be provided, if necessary.

1.10. MANAGEMENT SYSTEM

No additional information is needed for Groups 2 and 3.

Group 4: The management system should cover the following activities, if necessary:

(a) Package design, PDSR, documentation and records;
(b) Manufacture and testing of the packaging;
(c) Operation (i.e. preparation, loading, carriage, storage in transit, shipment after storage, unloading and receipt);
(d) Maintenance, repair and inspection of the packaging.

The management system should include descriptions of the actions to be performed to check compliance of the documents relating to package operation (e.g. manufacturing, operation or maintenance manuals) with the PDSR and should also cover the management of deviations detected in the framework of any transport activity.

For all components important to safety, the PDSR should define the parameters to be guaranteed to ensure compliance with the package design, and therefore safety, and the level of controls required during manufacturing and maintenance.

1.11. PACKAGE ILLUSTRATION

No additional information is needed for Group 2.

Group 3: A reproducible illustration should be provided showing the make-up of the package, including shock absorbers and internal arrangements, if applicable.

The illustration should indicate at least the overall outside dimensions and the mass of the package when empty and when loaded.

Group 4: A reproducible illustration should be provided showing the make-up of the package, including shock absorbers, devices for thermal protection, and internal arrangements, if applicable.

The illustration should indicate at least the overall outside dimensions and the mass of the package when empty and when loaded.

Part 2

2.1. STRUCTURAL ANALYSIS

There is no need for an additional structural analysis for Group 2.

An additional structural analysis should be performed under the following conditions:

— For Group 3, an additional structural analysis for industrial packages should be included if the demonstration of compliance with the requirements established in para. 674(b) or (c) of the Transport Regulations relies on the performance of the package under normal conditions of transport and is not otherwise assessed for the package design.

— For Group 4, an additional structural analysis should be included if the criticality safety assessment relies on the performance of the package under normal or accident conditions of transport and is not otherwise assessed for the package design.

The additional assessment of the mechanical behaviour (including, as applicable, analysis of thermal stresses, fatigue, brittle fracture and creep) for routine, normal and accident conditions of transport should include the following:

(a)   The components of the confinement system.
(b)   Any other package components (e.g. shock absorbing components, packaging components that provide heat dissipation) whose performance may have a consequential effect on (a).
(c)   The mechanical stability of the fissile material and any structure that is used to maintain its geometry, if necessary, for the criticality safety assessment. Other important criticality safety relevant items that should be considered are water leaking into or out of the package (totally or partially), the rearrangement of the fissile material and the degradation of neutron traps.

See also the remarks in item 1.5 of the PDSR.

When performing the structural analysis, the following points should be considered:

(1)   General considerations:
   (i)   The mechanical properties of the materials considered in the safety demonstration should be representative of the range of mechanical properties of the package components. The analysis should also consider the temperature range of the package components for an ambient temperature range of –40°C to +38°C or another temperature range specified by the competent authority in accordance with para. 679 of the Transport Regulations.
   (ii)   The impacts on the package behaviour due to variations in the shock absorbing properties of the shock absorber material (e.g. wood, polymers, plaster, concrete) should be analysed for an ambient temperature range of –40°C to +38°C or another temperature range specified by the competent authority in accordance with para. 679 of the Transport Regulations, and for the range of moisture likely to be encountered during transport.
   (iii) The safety against brittle fracture should be analysed at –40°C or another temperature specified by the competent authority in accordance

with para. 679 of the Transport Regulations, for those components of the confinement system made of potentially brittle materials (e.g. ferritic steels, cast iron).

(iv) The strength of lid bolts should be verified for all drop orientations.

(v) Any excursion of stress into the plastic domain should be avoided to the extent possible for confinement system components such as bolts and gasket seats, which would need additional complex proofs concerning the mechanics of the rupture or the maintenance of sufficient gasket seating.

(vi) Any possible damage of metallic seals after drops due to vibrations or sliding of the lid should be evaluated.

(vii) The internal components (e.g. content, basket, cage) should be assessed to verify that they are not liable to damage the confinement system. For the evaluation of the impact of internal components on the packaging lid, the maximum possible gap between these components and the lid before the drop should be considered.

(viii) The condition of the confinement system should be determined to demonstrate compliance with the specifications in item 2.3 of the PDSR for an ambient temperature range of $-40°C$ to $+38°C$ or another temperature range specified by the competent authority in accordance with para. 679 of the Transport Regulations.

(ix) Retention of sufficient thermal protection should be demonstrated, after the mechanical tests for accident conditions of transport, to guarantee the confinement system during the thermal test.

(x) The effect of the thermal test on the mechanical behaviour of the package (e.g. thermal stresses and strains, thermo-mechanical interactions between package components, interactions of the package components with contents) should be considered.

(xi) An appropriate water immersion test should be considered.

(2) Considerations for experimental mechanical testing:

(i) The package orientations should be determined in accordance with paras 722.4 and 727.5 of SSG-26 (Rev. 1) [2]. The orientations should maximize the loading of the package (in terms of stress, strain, acceleration and deformation), with consideration of the different package components (e.g. cask body, lid system, shock absorber) and of the protection objective (i.e. criticality safety). For package orientations, the following tests should be considered:

— Tests that maximize the stresses and acceleration (e.g. flat, slap down). The greater the impact area, the harder the impact, under the assumption of constant stiffness per unit area.

  — Tests that maximize the deformation (e.g. on corners, on edges). The smaller the impact area, the greater the crushing.

  — Tests that maximize the damage to orifices, notably by a puncture bar. The containment components in the orifices are often thin and more liable than the body of the packaging to be damaged by the bar.

  — Tests that maximize the risk of perforation by a puncture bar, possibly oblique. If the impacted package surface is oblique with respect to the puncture bar, the initial impact takes place on an edge of the puncture bar and the risk of perforation is much higher.

(ii) For reduced scale models, geometry and material properties similar to the original design, or conservative geometry and material properties, should be used.

(iii) The results of the drop test with reduced scale models should be assessed to guarantee that they cover, or are transferable to, the original design.

(iv) Drop tests performed with reduced scale models should be demonstrated as representative for the following parameters and components:

  — Drop heights: It might be necessary to increase the drop heights during testing to simulate the total potential energy that would have been received by the package at full scale. This should be considered for drop tests where the characteristic deformation of the structure is not negligible in comparison to the drop height.

  — Appropriate geometry scaling of all components (e.g. lids, nuts and bolts, grooves for the seals).

  — Metallic gaskets: The same design, same material and homothetic transformation with regard to elastic restitution should be selected.

  — O-rings: The selection should be based on the similarity of the useful elastic restitution, taking into account the compression set. The change of material properties with the temperature conditions should be considered.

  — The scaling of tightening torques for bolts of the reduced scale model should take into account the dispersion of friction conditions, the precision of torques and the technical limitations in an exact geometrical and physical scaling of the containment system and of the confinement system components.

  — Welding seams should be similar in the scale model and package design.

        — For reduced scale model drop testing with significant deformations of shock absorbers, the original package performance should be carefully justified.

(v)   The experimental mechanical tests should be conducted and reported in accordance with the management system. The test report should address the verification of the package before testing, the description of the test site, the equipment used for measurements and its calibration data, and the results of the measurements performed. This report should also contain pictures showing and explaining the conditions under which the tests were performed and their results.

(3)   Considerations for calculations:

(i)   See point (2)(i) above.

(ii)   Validated computer codes should be used. Input parameters (e.g. material laws, characteristic values, boundary conditions) should describe sufficiently and precisely the real technical and/or physical problems, and the use of these parameters should be justified.

(iii)   If uncertainties exist regarding important input parameters (e.g. material laws), conservative design calculations — including the possible range of material properties — should be performed.

(iv)   Data used (e.g. material laws, boundary conditions, load assumptions) and calculation results should be documented comprehensively.

2.2. THERMAL ANALYSIS

The temperature of the package components should be assessed in accident conditions of transport (see para. 659(b) of the Transport Regulations). The thermal analysis should include the thermal behaviour of the following components:

(a)   The components of the confinement system;

(b)   The package components whose performance will have a consequential effect on (a).

The following should be considered:

(1)   The effects of insolation over a period of 12 hours in accordance with para. 657 of the Transport Regulations.

(2)   The solar insolation before and after the thermal test, as defined in para. 728 of the Transport Regulations.

(3)   The justification for simplifying the assumptions (e.g. absence of trunnions) used for calculation under normal and accident conditions of transport.

(4)	The analysis of the package in accident conditions of transport under the position (i.e. horizontal or vertical) least resistant to thermal tests.

(5)	The value of the surface absorptivity coefficient of the package. The surface absorptivity coefficient should not be lower than 0.8 (see para. 728(a) of the Transport Regulations) during and after the thermal test to account for deposits on the package surface. The surface absorptivity coefficient should also not be lower than the possible maximum value of the emissivity coefficient in routine conditions of transport.

(6)	The evaluation of the minimum and maximum temperatures of the various components of the packaging, taking account of all the possible positions for the radioactive contents.

(7)	The profile of heat power based on the burnup distribution in irradiated fuels.

(8)	The justification that the concentration of oxygen present in the furnace is controlled and in conformity with that obtained in a hydrocarbon fuel–air fire, when the thermal test is conducted in a furnace and when it is noted that some package components burn. In addition, control of heat input should be considered thoroughly.

(9)	The influence of combustible materials that generate additional heat input and affect the fire duration beyond the thermal test.

(10)	The safety margins on temperature results derived using numerical modelling commensurate with the uncertainty associated with the numerical model.

(11)	Demonstration that the spare volume in the gasket grooves allows for gasket thermal expansion in accident conditions of transport, unless appropriate justification is provided.

2.3. CONTAINMENT DESIGN ANALYSIS

The extent of the additional information to be considered in the containment design analysis depends on the assumptions used to demonstrate criticality safety with regard to the fissile material that escapes from the package (see para. 685(c) of the Transport Regulations) and the quantity of water that leaks into or out of all void spaces of the package (see para. 680 of the Transport Regulations).

2.4. CRITICALITY SAFETY ANALYSIS

For packages designed to transport fissile material not excepted by paras 417, 674 or 675 of the Transport Regulations (i.e. packages in Group 4), assessment of criticality safety for routine, normal and accident conditions of transport is required to be performed, both for isolated packages and for arrays of packages.

The following, if applicable, should be considered in the criticality safety analysis:

(a) Contents:

    (i) All possible configurations with any possible geometrical and physical characteristics (e.g. dimensional tolerances, positions of the components, density of powders in normal or accident conditions of transport) within the range set by the description of the contents and the packaging (see items 1.3–1.6 of the PDSR).

    (ii) Materials with a hydrogen concentration greater than that of water that may be present in the package.

    (iii) Natural or depleted uranium that may be present in the package, with appropriate assumptions about quantities and localization.

(b) Configurations to be analysed:

    (i) If special features preventing water leakage are considered for the criticality safety analysis for a package in isolation (see para. 680(a) of the Transport Regulations), the criterion for water tightness to be defined by the package designer and accepted by the competent authority should be given and justified in the PDSR. This criterion should be set in such a way as to exclude ingress of an amount of water that could influence the criticality safety assessment. The testing conditions defined in para. 680 of the Transport Regulations should be taken into account.

    (ii) For packages to be transported by air, subcriticality of the isolated package should be assessed under conditions consistent with Type C package tests, assuming reflection by at least 20 cm of water but no water in-leakage. If the behaviour of the package under these conditions is not assessed, typical envelope configurations should be considered, such as the following:

        — The fissile material contained in one package (without consideration of water ingress from outside the package) in spherical shape reflected by 20 cm of water.

        — The fissile material contained in one package (without consideration of water from outside the package), pure or mixed with all or part of the moderating materials of the package, in spherical shape, surrounded by the reflecting materials (e.g. steel, lead) of the package and reflected by 20 cm of water.

    (iii) In modelling, all the elements of structures that could increase the neutron multiplication should be taken into account.

    (iv) The package designer should check the qualification of criticality calculation tools and should specify the critical experiments used for benchmarking of the calculation tools, which should be representative of the single package in isolation and the applicable arrays of packages. Special attention should be paid to systems (e.g. low-moderation systems, fuel assemblies) for which the qualification base is not really extended and for which it is desirable to use calculation models that are conservative in terms of calculation assumptions and provide margins to compensate for the lack of qualification, when applicable.

(v) When appropriate, the justifications should take into account all the possible ranges of the masses and moderations. Credible conditions of transport that might lead to preferential (heterogeneous) flooding of packages increasing the neutron multiplication should be considered.

(vi) For certain configurations where the interactions can be dominating, the impact of the variations in density of the fissile medium should be studied.

(vii) The heterogeneous nature of the fissile material during transport should be considered.

(viii) The assumption of a homogeneous enrichment equal to the average enrichment for uranium dioxide for boiling water reactor assemblies should be proven to be conservative, especially if the assembly geometry may change under the test conditions specified in the Transport Regulations.

(ix) For spent fuel initially containing plutonium, a conservative irradiation level that takes into account the possible evolution of reactivity during irradiation should be considered.

Further recommendations on criticality safety assessments are given in appendix VI to SSG-26 (Rev. 1) [2]. Further recommendations on the application of burnup credit in criticality safety assessments of spent nuclear fuel are provided in IAEA Safety Standards Series No. SSG-27 (Rev. 1), Criticality Safety in the Handling of Fissile Material [10]. Additional information can be found in ISO 27468:2011 [11] and in publications from the Expert Group on Burn-up Credit Criticality Safety of the OECD Nuclear Energy Agency.[3]

---

[3] See https://www.oecd-nea.org/science/wpncs/buc/index.html

Appendix VI

**ADDITIONAL INFORMATION FOR PACKAGES CONTAINING
0.1 kg OR MORE OF URANIUM HEXAFLUORIDE**

VI.1. This appendix provides specific additional recommendations on the information to be provided in Parts 1 and 2 of the PDSR for packages containing 0.1 kg or more of uranium hexafluoride. Table 6 lists each item of the PDSR, with applicable information and guidance.

VI.2. The recommendations provided in this appendix apply in addition to those associated with the package type defined by the radioactive properties of the contents (see Appendices II–IV).

VI.3. Further recommendations are provided in SSG-26 (Rev. 1) [2].

VI.4. For packages containing fissile nuclides, see in addition Appendix V.

VI.5. For technical analyses, compliance of the package with the Transport Regulations should be demonstrated by the package designer either directly in item 1.6 of the PDSR or, if necessary, in Part 2 of the PDSR.

VI.6. For each of the technical analyses in Part 2 of the PDSR, the following considerations apply:

(a) The package design being evaluated should be uniquely identified by precisely indicating a drawing of the packaging (see item 1.3 of the PDSR), including its revision number, and the specification of the contents (see item 1.2 of the PDSR), including its revision number.
(b) The acceptance criteria for the technical analyses and the package design assumptions relating to geometry or performance characteristics should be defined and justified when necessary. The acceptance criteria should be specified by the designer and should be derived from the criteria established by the regulatory body and from other applicable standards developed to meet the regulatory requirements. The design assumptions refer to the design specifications provided in items 1.2 and 1.3 of the PDSR and to other assumptions derived from the design specifications and used in the technical analyses. All mechanical and thermal characteristics of each component of the package and the acceptance criteria used in technical analyses should be defined. The design assumptions should take into account ageing

77

mechanisms, as necessary. Further recommendations are provided in paras 613A.5 and 613A.6 of SSG-26 (Rev. 1) [2].

(c)   The compliance of a package design containing 0.1 kg or more of uranium hexafluoride is required to be accomplished in accordance with para. 701 of the Transport Regulations by any of the following methods or by a combination thereof:

    (i)   The results of physical testing of prototypes or models of appropriate scale. When a programme for physical testing of prototypes or models of appropriate scale is implemented for a specific package design to be approved by the competent authority, the competent authority should be notified of the programme before the testing and should be allowed to witness the testing. When such a programme of tests is done without competent authority approval but is part of the safety analysis, its validity is to be determined by the competent authority.

    (ii)   Reference to previous satisfactory demonstrations of a sufficiently similar nature. Test results of designs similar to the design under consideration are permissible if the similarity can be demonstrated sufficiently by justification and validation.

    (iii)   Calculation or reasoned argument, when the calculation procedures are generally agreed to be suitable and conservative. Assumptions should be clearly stated and fully justified, including by physical testing if applicable.

Further recommendations are provided in paras 701.1–701.25 of SSG-26 (Rev. 1) [2].

The methods or standards used in each analysis specified in items 2.1–2.3 of the PDSR should include a description of the analysis technique used, the limitations and accuracy of this technique and a demonstration of the correct application of the technique for the analysis of the package design.

If computer codes are used for the safety analysis, additional information should be included in the PDSR to demonstrate that the code is verified and validated for the specific field of use. The justification for the applicability of these codes should include a statement of possible sources of error and/or uncertainty relative to the effects of the operating platform (computer) used and of modelling assumptions and simplifications, as well as of any other parameter influencing the calculated results and a sensitivity analysis.

(d)   The performance characteristics of the package design should be assessed. More than one sequence of tests might need to be considered to ensure that the various safety functions to be fulfilled by different components of the package design comply with the regulatory requirements. Other hazards (e.g. corrosion, combustion, pyrophoricity or other chemical reactions, radiolysis, phase changes) should be analysed, as necessary, if they have a consequential effect on the safety functions of the package.

(e) The results of the technical analyses should be compared with the acceptance criteria, and package design assumptions and regulatory compliance should be justified accordingly.

VI.7. The following items listed in Fig. 1 and para. 2.3 of this Safety Guide are not relevant to and not needed for packages containing 0.1 kg or more of uranium hexafluoride and, therefore, are not included in Table 6:

— Part 2:
  - Dose rate analysis;
  - Criticality safety analysis;
  - Other analyses.

APPENDIX VI
TABLE 6. PACKAGE DESIGN SAFETY REPORT FOR PACKAGES
CONTAINING 0.1 kg OR MORE OF URANIUM HEXAFLUORIDE,
ADDITIONAL INFORMATION

Part 1

1.1. ADMINISTRATIVE INFORMATION

The colloquial name of the package, if applicable, should be added.

1.2. SPECIFICATION OF THE CONTENTS

A detailed description of the permitted contents in accordance with the requirements for packages containing uranium hexafluoride established in para. 420 of the Transport Regulations should be provided.

1.3. SPECIFICATION OF THE PACKAGING

The following information should be added, when necessary:

(a) A list of packaging components important to safety and their materials, including the specifications and methods of manufacture of the components, specifications for material procurement, welding, other special processes, and non-destructive examination and testing. The properties of materials of components expected to maintain their safety function under the thermal test should be given for a range of temperature reachable during such a test.
(b) A description of the packaging body and closure mechanism.
(c) A description of the packaging components for thermal protection.
(d) Testing specifications and controls before first use to transport uranium hexafluoride. This ensures compliance of the fabrication with the design and allows acceptance of the specimen before its first use. See also para. 501 of the Transport Regulations.

1.4. AGEING CONSIDERATIONS

Depending on the package design, the information relating to ageing considerations can also be provided by the package designer directly in the table mentioned in item 1.6.

For packaging used once for a single transport and not intended for shipment after storage, this item of the PDSR should be left blank.

For all other packaging, this item of the PDSR should include the following information:

(a) The intended conditions of use of the package that might influence ageing;
(b) The potential ageing mechanisms relevant to the package design, taking into account the intended conditions of use of the package;
(c) Operational measures (including maintenance and inspection before shipment) to monitor and limit the ageing effects;
(d) Analysis of the influence of ageing of the packaging and contents on the design assumptions for demonstration of compliance with the regulations, including the technical analyses in Part 2 of the PDSR, considering the specified intended use conditions, ageing mechanisms and operational measures.

Further recommendations are provided in paras 613A.5 and 613A.6 of SSG-26 (Rev. 1) [2].

1.5. CONDITIONS FOR TECHNICAL ANALYSES

This item of the PDSR should describe the main principles and performance characteristics of the package design to meet the safety requirements (e.g. leakage, stress, rupture) of the Transport Regulations.

This item should summarize the analyses performed in Part 2 of the PDSR and describe how analysis assumptions and data used for the safety analysis — especially regarding leakage, stress and rupture — are derived from the design and the behaviour of the package for routine, normal and accident conditions of transport, also taking into account ageing mechanisms (see item 1.4 of the PDSR).

This item will help ensure that the package design and the various parts of the safety demonstration are compatible with one another.

1.6. COMPLIANCE WITH REGULATORY REQUIREMENTS

The PDSR should include a complete list of all paragraphs of the Transport Regulations and other international or national regulations applicable to packages containing 0.1 kg or more of uranium hexafluoride.

Compliance of the package design with these regulations should be demonstrated in the PDSR. This could be demonstrated using a table (or any other written format) linking the appropriate items of the PDSR, where compliance is demonstrated, to the applicable paragraphs of the regulations.

The applicable paragraphs of the Transport Regulations for packages containing 0.1 kg or more of uranium hexafluoride are provided in a matrix in Annex I.

1.7. PACKAGE OPERATIONS

The PDSR should include the measures to be implemented to ensure compliance with the requirements established in para. 420 of the Transport Regulations.

1.8. MAINTENANCE

The PDSR should include the measures to be implemented to ensure compliance with ISO 7195:2020 [12], as required by para. 631 of the Transport Regulations.

1.9. GAP ANALYSIS PROGRAMME

For packages that are to be used for shipment after storage, the PDSR should include a gap analysis programme describing a systematic procedure for a periodic evaluation of changes of regulations, changes in technical knowledge and changes in the state of the package design during storage (see also paras 613A.5, 809.3 and 809.4 of SSG-26 (Rev. 1) [2]).

1.10. MANAGEMENT SYSTEM

The PDSR should specify the management system established and implemented by the package designer, in accordance with para. 306 of the Transport Regulations, to demonstrate compliance with the relevant provisions of the Transport Regulations.

The management system should cover the following activities:

(a) Package design, PDSR, documentation and records;
(b) Manufacture and testing of the packaging;
(c) Operation (i.e. preparation, loading, carriage, storage in transit, shipment after storage, unloading and receipt);
(d) Maintenance, repair and inspection of the packaging.

The management system should be commensurate with the complexity of the package design and should include a reliable document control system.

The management system should include descriptions of the actions to be performed to check compliance of the documents relating to package operation (e.g. manufacturing, operation or maintenance manuals) with the PDSR and should also cover the management of deviations detected in the framework of any transport activity.

For all components important to safety, the PDSR should define the parameters to be guaranteed and the level of controls required during manufacturing and maintenance.

Further recommendations on the management system for transport are provided in TS-G-1.4 [6].

1.11. PACKAGE ILLUSTRATION

A reproducible illustration should be provided showing the make-up of the package, including shock absorbers and devices for thermal protection, if applicable.

The illustration should indicate at least the overall outside dimensions and the mass of the package when empty and when loaded.

Part 2

2.1. STRUCTURAL ANALYSIS

The general considerations for all technical analyses should be taken into account when performing the structural analysis to demonstrate compliance with the requirements established in para. 632(a) and (b) of the Transport Regulations.

The assessment of the mechanical behaviour (including, as applicable, analysis of thermal stresses, fatigue, brittle fracture and creep) for routine, normal and accident conditions of transport should include the following:

(a) The components of the containment system;
(b) Any other package components (e.g. shock absorbing components, packaging components that provide thermal protection) whose performance may have a consequential effect on (a).

When performing the structural analysis for packages containing 0.1 kg or more of uranium hexafluoride, the following points should be considered:

(1) General considerations:
  (i) The mechanical properties of the materials considered in the safety demonstration should be representative of the range of mechanical properties of the package components, with account taken of the temperature range applicable to the type of package.
  (ii) The impacts on the package behaviour due to variations in the shock absorbing properties of the shock absorber material should be analysed, considering the temperature range applicable to the type of package and the range of moisture conditions likely to be encountered during transport.

(iii)   The safety against brittle fracture should be analysed, considering the temperature range applicable to the type of package.

(iv)   It should be verified that the content is not liable to damage the containment system.

(v)   The condition of the containment system should be assessed to demonstrate compliance with the specifications in item 2.3 of the PDSR for the temperature range applicable to the type of package.

(vi)   The ability to withstand the maximum pressure during the thermal test (elevation of pressure of uranium hexafluoride) should be demonstrated.

(2)   Considerations for experimental mechanical testing:

(i)   The package orientations should be determined in accordance with para. 722.4 of SSG-26 (Rev. 1) [2]. The orientations should maximize the loading of the package (in terms of stress, strain, acceleration and deformation), with consideration of the different package components (e.g. cylinder body, shock absorber) and of the protection objective (i.e. containment). For package orientations, the following tests should be considered, as appropriate:

— Tests that maximize the stresses and acceleration (e.g. flat, slap down). The greater the impact area, the harder the impact, under the assumption of constant stiffness per unit area.

— Tests that maximize the deformation (e.g. on corners, on edges). The smaller the impact area, the greater the crushing.

— Tests that maximize the damage to valves, notably by a puncture bar.

— Tests that maximize the risk of perforation by a puncture bar, possibly oblique. If the impacted package surface is oblique with respect to the puncture bar, the initial impact takes place on an edge of the puncture bar and the risk of perforation is much higher.

(ii)   For reduced scale models, geometry and material properties similar to the original design, or conservative geometry and material properties, should be used.

(iii)   The results of the drop test with reduced scale models should be assessed to guarantee that they cover, or are transferable to, the original design.

(iv)   Drop tests performed with reduced scale models should be demonstrated as representative for the following parameters and components:

— Drop heights: It might be necessary to increase the drop heights during testing to simulate the total potential energy that would have been received by the package at full scale. This should be considered for drop tests where the characteristic deformation of the structure is not negligible in comparison to the drop height.

— Appropriate geometry scaling of all components (e.g. lids, nuts and bolts, grooves for the seals, valves, plugs).

— The scaling of tightening torques for bolts of the reduced scale model should take into account the dispersion of friction conditions, the precision of torques and the technical limitations in an exact geometrical and physical scaling of the containment system components.

— Welding seams should be similar for scale model and package design.

— For reduced scale model drop testing with significant deformations of shock absorbers, the original package performance should be carefully justified.

(v) The experimental mechanical tests should be conducted and reported in accordance with the management system. The test report should address the verification of the package before testing, the description of the test site, the equipment used for measurements and its calibration data, and the results of the measurements performed. This report should also contain pictures showing and explaining the conditions under which the tests were performed and their results.

(3) Considerations for calculations:

(i) See point (2)(i) above.

(ii) Validated computer codes should be used. Input parameters (e.g. material laws, characteristic values, boundary conditions) should describe sufficiently and precisely the real technical and/or physical problems, and the use of these parameters should be justified.

(iii) If uncertainties exist regarding important input parameters (e.g. material laws), conservative design calculations — including the possible range of material properties — should be performed.

(iv) Data used (e.g. material laws, boundary conditions, load assumptions) and calculation results should be documented comprehensively.

2.2. THERMAL ANALYSIS

The temperature of the package components should be assessed in accident conditions of transport (see para. 659(b) of the Transport Regulations). The thermal analysis should include the thermal behaviour of the following components:

(a) The components of the containment system;

(b) The package components whose performance will have a consequential effect on (a).

The demonstration of compliance should be limited to showing compliance with para. 632(c) of the Transport Regulations.

The following should be considered:

(1)	The effects of insolation over a period of 12 hours in accordance with para. 657 of the Transport Regulations.
(2)	The solar insolation before and after the thermal test, as defined in para. 728 of the Transport Regulations.
(3)	The justification for simplifying the assumptions (e.g. absence of trunnions) used for calculation under normal and accident conditions of transport.
(4)	The analysis of the package in accident conditions of transport in the position (i.e. horizontal or vertical) least resistant to thermal tests.
(5)	The value of the surface absorptivity coefficient of the package. The surface absorptivity coefficient should not be lower than 0.8 (see para. 728(a) of the Transport Regulations) during and after the thermal test to account for deposits on the package surface. The surface absorptivity coefficient should also not be lower than the possible maximum value of the emissivity coefficient in routine conditions of transport.
(6)	The evaluation of the minimum and maximum temperatures of the various components of the packaging, taking account of all the possible positions for the radioactive contents.
(7)	The justification that the concentration of oxygen present in the furnace is controlled and in conformity with that obtained in a hydrocarbon fuel–air fire, when the thermal test is conducted in a furnace and when it is noted that some package components burn. In addition, control of heat input should be considered thoroughly.
(8)	The influence of combustible materials that generate additional heat input and affect the fire duration beyond the thermal test.
(9)	The safety margins on temperature results derived using numerical modelling commensurate with the uncertainty associated with the numerical model.

2.2. CONTAINMENT DESIGN ANALYSIS

Compliance with para. 632(a) and (b) of the Transport Regulations should be demonstrated in the structural analysis. Compliance with para. 632(c) of the Transport Regulations should be demonstrated in the thermal analysis.

REFERENCES

[1]   INTERNATIONAL ATOMIC ENERGY AGENCY, Regulations for the Safe Transport of Radioactive Material, 2018 Edition, IAEA Safety Standards Series No. SSR-6 (Rev. 1), IAEA, Vienna (2018).

[2]   INTERNATIONAL ATOMIC ENERGY AGENCY, Advisory Material for the IAEA Regulations for the Safe Transport of Radioactive Material (2018 Edition), IAEA Safety Standards Series No. SSG-26 (Rev. 1), IAEA, Vienna (in preparation).

[3]   UNITED NATIONS, Recommendations on the Transport of Dangerous Goods: Model Regulations — Twenty-First Revised Edition, ST/SG/AC.10/1/Rev.21, 2 vols, United Nations, New York and Geneva (2019).

[4]   EUROPEAN ASSOCIATION OF COMPETENT AUTHORITIES, Package Design Safety Reports for the Transport of Radioactive Material, European PDSR Guide Issue 3 (December 2014), EACA (2014).

[5]   INTERNATIONAL ATOMIC ENERGY AGENCY, IAEA Safety Glossary: Terminology Used in Nuclear Safety and Radiation Protection, 2018 Edition, IAEA, Vienna (2019).

[6]   INTERNATIONAL ATOMIC ENERGY AGENCY, The Management System for the Safe Transport of Radioactive Material, IAEA Safety Standards Series No. TS-G-1.4, IAEA, Vienna (2008).

[7]   INTERNATIONAL COMMISSION ON RADIOLOGICAL PROTECTION, Nuclear Decay Data for Dosimetric Calculations, ICRP Publication 107, Ann. ICRP 38 (3), Elsevier (2008).

[8]   INTERNATIONAL ATOMIC ENERGY AGENCY, Methodology for a Safety Case of a Dual Purpose Cask for Storage and Transport of Spent Fuel, IAEA-TECDOC-1938, IAEA, Vienna (2020).

[9]   INTERNATIONAL ATOMIC ENERGY AGENCY, Application of the Revised Provisions for Transport of Fissile Material in the IAEA Regulations for the Safe Transport of Radioactive Material, 2012 Edition, IAEA-TECDOC-1768, IAEA, Vienna (2015).

[10]   INTERNATIONAL ATOMIC ENERGY AGENCY, Criticality Safety in the Handling of Fissile Material, IAEA Safety Standards Series No. SSG-27 (Rev. 1), IAEA, Vienna (in preparation).

[11]   INTERNATIONAL ORGANIZATION FOR STANDARDIZATION, Nuclear Criticality Safety — Evaluation of Systems Containing PWR UOX Fuels — Bounding Burnup Credit Approach, ISO 27468:2011, ISO, Geneva (2011).

[12]   INTERNATIONAL ORGANIZATION FOR STANDARDIZATION, Nuclear Energy — Packagings of Uranium Hexafluoride (UF6) for Transport, ISO 7195:2020, ISO, Geneva (2020).

MATRIX OF APPLICABLE REQUIREMENTS OF THE TRANSPORT REGULATIONS FOR DIFFERENT PACKAGE TYPES AND ADDITIONAL PROVISIONS

I–1. Table I–1 shows the applicable paragraphs of the Transport Regulations to be included in the demonstration of compliance for each package type and additional provisions for packages containing fissile material and packages containing 0.1 kg or more of uranium hexafluoride.

TABLE I–1. MATRIX OF APPLICABLE REQUIREMENTS OF THE TRANSPORT REGULATIONS FOR DIFFERENT PACKAGE TYPES AND ADDITIONAL PROVISIONS

Paras of SSR-6 (Rev. 1)[a]	Package type						Additional provisions			Remarks
	Excepted	IP-1	IP-2	IP-3	A	B(U), B(M)	C	Fissile	UF$_6$	
Activity limits and classification										
408–411		X	X	X						Low specific activity (LSA) classification and activity limits Paragraph 410: transport by air
412–414		X	X	X						Surface contaminated object (SCO) classification and activity limits
415, 416	X				X	X				If special form radioactive material or low dispersible radioactive material is present, add reference to approval certificate
417, 418								X		Classification as fissile material, exceptions and restrictions
419, 420									X	Classification as uranium hexafluoride and restrictions
422–427	X									Classification as excepted package

TABLE I–1. MATRIX OF APPLICABLE REQUIREMENTS OF THE TRANSPORT REGULATIONS FOR DIFFERENT PACKAGE TYPES AND ADDITIONAL PROVISIONS (cont.)

Paras of SSR-6 (Rev. 1)[a]	Package type						Additional provisions			Remarks
	Excepted	IP-1	IP-2	IP-3	A	B(U), B(M)	C	Fissile	UF_6	
429, 430					X					Activity limit for Type A package
431, 432							X			Classification as Type C package and activity limits
433						X				Activity limits for Type B(U) and Type B(M) packages by air
Requirements and controls for transport										
501		X	X	X	X	X	X			Requirements before first shipment
502, 503	X	X	X	X	X	X	X			Requirements before each shipment
504		X	X	X	X	X	X			Additional items in the package
507	X	X	X	X	X	X	X			Other dangerous properties
515, 516	X									Excepted package requirements

TABLE I–1. MATRIX OF APPLICABLE REQUIREMENTS OF THE TRANSPORT REGULATIONS FOR DIFFERENT PACKAGE TYPES AND ADDITIONAL PROVISIONS (cont.)

Paras of SSR-6 (Rev. 1)[a]	Package type						Additional provisions			Remarks
	Excepted	IP-1	IP-2	IP-3	A	B(U), B(M)	C	Fissile	UF$_6$	
517		X	X	X						Dose rate of unshielded LSA or SCO
521		X	X	X						Package for LSA material and SCO
522		X	X	X						Activity limit for LSA and SCO
526		X	X	X	X	X	X	X		Transport index and criticality safety index (CSI) limits
527, 528		X	X	X	X	X	X			Dose rate at contact of a package
533	X	X	X	X	X	X	X			
534		X	X	X						Marking
535, 536						X	X	X	X	
575		X	X	X	X	X	X			Transport by sea
578						X				Transport by air for Type B(M) package

TABLE I–1. MATRIX OF APPLICABLE REQUIREMENTS OF THE TRANSPORT REGULATIONS FOR DIFFERENT PACKAGE TYPES AND ADDITIONAL PROVISIONS (cont.)

Paras of SSR-6 (Rev. 1)[a]	Package type							Additional provisions		Remarks
	Excepted	IP-1	IP-2	IP-3	A	B(U), B(M)	C	Fissile	UF_6	
Requirements for radioactive material and for packagings and packages										
606	X	X	X	X	X	X	X			Requirements for fissile excepted classification
607–618	X	X	X	X	X	X	X			General provisions
619–621	X	X	X	X	X	X	X			Transport by air
622	X									Excepted packages
623		X								Type IP-1
624			X							Type IP-2
625				X						Type IP-3
626			X							Alternative requirements for Type IP-2

TABLE I–1. MATRIX OF APPLICABLE REQUIREMENTS OF THE TRANSPORT REGULATIONS FOR DIFFERENT PACKAGE TYPES AND ADDITIONAL PROVISIONS (cont.)

Paras of SSR-6 (Rev. 1)[a]	Package type						Additional provisions			Remarks
	Excepted	IP-1	IP-2	IP-3	A	B(U), B(M)	C	Fissile	UF$_6$	
627–630			X	X						Alternative requirements
631–634									X	Uranium hexafluoride
636		X	X	X	X	X	X	X		Minimum dimensions
637			X	X	X	X	X			Seal
638–647			X	X	X	X				Type A
648			X	X	para. 648(b) only	para. 648(b) only				Type A, release criteria
649			X	X	X	X				Type A, liquids
650				X						Type A, liquids
651				X						Type A, gases

TABLE I–1. MATRIX OF APPLICABLE REQUIREMENTS OF THE TRANSPORT REGULATIONS FOR DIFFERENT PACKAGE TYPES AND ADDITIONAL PROVISIONS (cont.)

Paras of SSR-6 (Rev. 1)[a]	Package type						Additional provisions			Remarks
	Excepted	IP-1	IP-2	IP-3	A	B(U), B(M)	C	Fissile	UF$_6$	
652						X				Type B(U)
653–657						X	X			Type B(U)
658–660						X				Type B(U)
661–666						X	X			Type B(U)
667, 668						X				Type B(M) only
669–672							X			Type C
673								X		Fissile material
674, 675								X		CSI control with exception from competent authority approval of package design for fissile material
676–686								X		Packages containing fissile material

TABLE I–1. MATRIX OF APPLICABLE REQUIREMENTS OF THE TRANSPORT REGULATIONS FOR DIFFERENT PACKAGE TYPES AND ADDITIONAL PROVISIONS (cont.)

| Paras of SSR-6 (Rev. 1)[a] | Package type | | | | | | Additional provisions | | | Remarks |
	Excepted	IP-1	IP-2	IP-3	A	B(U), B(M)	C	Fissile	UF_6	
Test procedures										
701	X	X	X	X	X	X	X	X	X	Demonstration of compliance
702			X	X	X	X	X	X	X	Assessment after tests
713–715			X	X	X	X	X	X	X	Preparation of a package for testing
716			X	X	X	X	X	X	X	Integrity of containment and shielding, assessing criticality safety
717			X	X	X	X	X	X	X	Target for drop tests
718									X	Structural test
719, 720				X	X	X	X	X	X	General provisions for normal conditions tests
721				X	X	X	X	X		Water spray test

TABLE I–1. MATRIX OF APPLICABLE REQUIREMENTS OF THE TRANSPORT REGULATIONS FOR DIFFERENT PACKAGE TYPES AND ADDITIONAL PROVISIONS (cont.)

Paras of SSR-6 (Rev. 1)[a]	Package type						Additional provisions			Remarks
	Excepted	IP-1	IP-2	IP-3	A	B(U), B(M)	C	Fissile	UF$_6$	
722			X	X	X	X	X	X	X	Free drop test
723			X	X	X	X	X	X		Stacking test
724				X	X	X	X	X		Penetration test
725					X					Additional tests for Type A (liquids and gases)
726						X	X	X		General provisions for accident conditions tests
727(a)						X	X	X		9 m drop test
727(b)						X		X		Drop test onto a bar
727(c)						X	X	X		Dynamic crush test
728						X		X	X	Thermal test

TABLE I–1. MATRIX OF APPLICABLE REQUIREMENTS OF THE TRANSPORT REGULATIONS FOR DIFFERENT PACKAGE TYPES AND ADDITIONAL PROVISIONS (cont.)

Paras of SSR-6 (Rev. 1)[a]	Package type						Additional provisions			Remarks
	Excepted	IP-1	IP-2	IP-3	A	B(U), B(M)	C	Fissile	UF$_6$	
729						X		X		Water immersion test
730						X	X			Enhanced water immersion test
731–733								X		Water leakage test
734							X			General provisions for Type C package tests
735							X			Puncture–tearing test
736							X			Enhanced thermal test
737							X			Impact test

[a] INTERNATIONAL ATOMIC ENERGY AGENCY, Regulations for the Safe Transport of Radioactive Material, 2018 Edition, IAEA Safety Standards Series No. SSR-6 (Rev. 1), IAEA, Vienna (2018).

Annex II

**REFERENCE PUBLICATIONS USED BY COMPETENT
AUTHORITIES FOR TECHNICAL ASSESSMENTS**

II–1. Table II–1 provides a list of reference publications used by different competent authorities for the technical assessment of package designs.

TABLE II–1. REFERENCE DOCUMENTS USED BY COMPETENT AUTHORITIES FOR TECHNICAL ASSESSMENTS

Canada	ISO 2919, Radiological Protection — Sealed Radioactive Sources — General Requirements and Classification
	ISO 9978, Radiation Protection — Sealed Sources — Leakage Test Methods
	ISO 7195, Nuclear Energy — Packagings of Uranium Hexafluoride (UF6) for Transport
	ANSI N14.1, Uranium Hexafluoride — Packaging for Transport
	ISO 12807, Safe Transport of Radioactive Materials — Leakage Testing on Packages
	ANSI N14.5, Leakage Tests on Packages for Shipment
	ANSI N14.7, Guidance for Packaging Type A Quantities of Radioactive Materials
	RD-364, Joint Canada–United States Guide for Approval of Type B(U) and Fissile Material Transportation Packages
	ISO 9001, Quality Management Systems — Requirements
France	ASN Guide No. 7, Transport — Applicant's guide related to applications for shipment approval and certificate of package design or radioactive materials for civil usage transported by public roads, by water or by railroad
	ISO 2919, Radiological Protection — Sealed Radioactive Sources — General Requirements and Classification
	ISO 9978, Radiation Protection — Sealed Sources — Leakage Test Methods
	ISO 7195, Nuclear Energy — Packagings of Uranium Hexafluoride (UF6) for Transport
	ANSI N14.1, Uranium Hexafluoride — Packaging for Transport
	ISO 12807, Safe Transport of Radioactive Materials — Leakage Testing on Packages
	ISO 10276, Nuclear Energy — Fuel Technology for Packages Used to Transport Radioactive Material
	NF EN 25-030-1, Éléments de fixation — Assemblages vissés — Partie 1: règles générales de conception, de calcul et de montage (2014)
	VDI 2230, Systematic Calculation of High Duty Bolted Joints
	YOUNG, W.C., BUOYNAS, R.E., ROARK's Formulas for Stress and Strain, 7th edn, McGraw-Hill, New York (1984)

France (cont.)	CEA, Catalogue PMDS, Tome I, Ecrans de protection contre les rayonnements ionisants
	NF EN 10228-3, Essais non destructifs des pièces forgées en acier — contrôle par ultrasons des pièces forgées en aciers ferritiques et martensitiques
Germany	ANSI N14.1, Uranium Hexafluoride — Packaging for Transport
	BAM-GGR 007, Leitlinie zur Verwendung von Gusseisen mit Kugelgraphit für Transport- und Lagerbehälter für radioaktive Stoffe
	BAM-GGR 008, Richtlinie für numerisch geführte Sicherheitsnachweise im Rahmen der Bauartprüfung von Transport- und Lagerbehältern für radioaktive Stoffe
	BAM-GGR 011, Quality Assurance Measures of Packagings for Competent Authority Approved Package Designs for the Transport of Radioactive Material
	BAM-GGR 012, Leitlinie zur Berechnung der Deckelsysteme und Lastanschlagsysteme von Transportbehältern für radioaktive Stoffe
	DIN 25415 part 1, Radioactively Contaminated Surfaces — Method for Testing and Assessing the Ease of Decontamination
	FKM, Guideline "Fracture Mechanics Proof of Strength for Engineering Components"
	FKM, Richtlinie „Rechnerischer Festigkeitsnachweis für Maschinenbauteile"
	ISO 2919, Radiological Protection — Sealed Radioactive Sources General Requirements and Classification
	ISO 7195, Nuclear Energy — Packagings of Uranium Hexafluoride (UF6) for Transport
	ISO 9978, Radiation Protection — Sealed Sources — Leakage Test Methods
	ISO 12807, Safe Transport of Radioactive Materials — Leakage Testing on Packages
	KTA 3905, Load Attachment Points on Loads in Nuclear Power Plants
	BRITISH ENERGY GENERATION, R6 — Assessment of the Integrity of Structures Containing Defects
	VDI 2230, Systematic Calculation of High Duty Bolted Joints
	DIN 25712, Criticality Safety Taking into Account the Burnup of Fuel for Transport and Storage of Irradiated Light Water Reactor Fuel Assemblies in Casks
	ICRP Publication 103, The 2007 Recommendations of the International Commission on Radiological Protection
	ICRP Publication 74, Conversion Coefficients for Use in Radiological Protection against External Radiation
United States of America	ASME Boiler and Pressure Vessel Code, Section III, Division 3 — Containment Systems and Transport Packagings for Spent Nuclear Fuel and High-Level Radioactive Waste (2015 or later editions)

TABLE II–1. REFERENCE DOCUMENTS USED BY COMPETENT
AUTHORITIES FOR TECHNICAL ASSESSMENTS (cont.)

United States of America (cont.)	ASME Boiler and Pressure Vessel Code, Section III, Division 1 — Containment Systems and Transport Packagings for Spent Nuclear Fuel and High-Level Radioactive Waste (2015 or later editions)
	ANSI/ANS 6.1.1, American National Standard for Neutron and Gamma-Ray Fluence-to-Dose Factors (1977)
	ANSI N14.5-2014, American National Standard for Radioactive Materials — Leakage Tests on Packages for Shipment
	ANSI N14.1-2019, Uranium Hexafluoride Packagings Transport
	NRC Draft Regulatory Issue Summary NRC-2015-0047, Considerations in Licensing High Burnup Spent Fuel in Dry Storage and Transportation
	NRC Regulatory Guide 7.9, Standard Format and Content of Part 71 Applications for Approval of Packages for Radioactive Material, revision 2 (2005)
	NRC Interim Staff Guidance 8, Burnup Credit in the Criticality Safety Analyses of PWR Spent Fuel in Transportation and Storage Casks, revision 3
	NUREG-2216, Standard Review Plan for Transportation Packages for Spent Fuel and Radioactive Material
	NUREG/CR-5661, Recommendations for Preparing the Criticality Safety Evaluation of Transportation Packages
	NUREG/CR-6361, Criticality Benchmark Guide for Light-Water-Reactor Fuel in Transportation and Storage Packages
	NUREG/CR-6802, Recommendations for Shielding Evaluations for Transport and Storage Packages
	NUREG/CR-6487, Containment Analysis for Type B Packages Used to Transport Various Contents

Annex III

STRUCTURE OF THE PACKAGE DESIGN SAFETY
REPORT FOR APPENDICES I–VI

Table III–1 shows the structure of the PDSR for the packages addressed in Appendices I–VI of this Safety Guide.

TABLE III–1. STRUCTURE OF THE PACKAGE DESIGN SAFETY REPORT FOR APPENDICES I–VI

APPENDIX I TABLE 1: Excepted packages		APPENDIX II TABLE 2: Industrial packages		APPENDIX III TABLE 3: Type A packages		APPENDIX IV TABLE 4: Type B(U), Type B(M) and Type C packages		APPENDIX V TABLE 5: Packages containing fissile nuclides, additional information		APPENDIX VI TABLE 6: Packages containing uranium hexafluoride, additional information	
Part 1		**Part 1**		**Part 1**		**Part 1**		**Part 1**		**Part 1**	
1.1.	Administrative information	1.1.	Administrative information	1.1.	Administrative information	1.1.	Administrative information	1.1.	Administrative information	1.1.	Administrative information
1.2.	Specification of the contents	1.2.	Specification of the contents	1.2.	Specification of the contents	1.2.	Specification of the contents	1.2.	Specification of the contents	1.2.	Specification of the contents
1.3.	Specification of the packaging	1.3.	Specification of the packaging	1.3.	Specification of the packaging	1.3.	Specification of the packaging	1.3.	Specification of the packaging	1.3.	Specification of the packaging
		1.4.	Ageing considerations	1.4.	Ageing considerations	1.4.	Ageing considerations	1.4.	Ageing considerations	1.4.	Ageing considerations
						1.5.	Conditions for technical analyses	1.5.	Conditions for technical analyses	1.5.	Conditions for technical analyses

TABLE III–1. STRUCTURE OF THE PACKAGE DESIGN SAFETY REPORT FOR APPENDICES I–VI (cont.)

APPENDIX I TABLE 1: Excepted packages		APPENDIX II TABLE 2: Industrial packages		APPENDIX III TABLE 3: Type A packages		APPENDIX IV TABLE 4: Type B(U), Type B(M) and Type C packages		APPENDIX V TABLE 5: Packages containing fissile nuclides, additional information		APPENDIX VI TABLE 6: Packages containing uranium hexafluoride, additional information	
1.4.	Compliance with regulatory requirements	1.5.	Compliance with regulatory requirements	1.5.	Compliance with regulatory requirements	1.6.	Compliance with regulatory requirements	1.6.	Compliance with regulatory requirements	1.6.	Compliance with regulatory requirements
1.5.	Package operations	1.6.	Package operations	1.6.	Package operations	1.7.	Package operations	1.7.	Package operations	1.7.	Package operations
1.6.	Maintenance	1.7.	Maintenance	1.7.	Maintenance	1.8.	Maintenance	1.8.	Maintenance	1.8.	Maintenance
		1.8.	Gap analysis programme	1.8.	Gap analysis programme	1.9.	Gap analysis programme	1.9.	Gap analysis programme	1.9.	Gap analysis programme
1.7.	Management system	1.9.	Management system	1.9.	Management system	1.10.	Management system	1.10.	Management system	1.10.	Management system
				1.10.	Package illustration	1.11.	Package illustration	1.11.	Package illustration	1.11.	Package illustration

TABLE III–1. STRUCTURE OF THE PACKAGE DESIGN SAFETY REPORT FOR APPENDICES I–VI (cont.)

APPENDIX I TABLE 1: Excepted packages		APPENDIX II TABLE 2: Industrial packages		APPENDIX III TABLE 3: Type A packages		APPENDIX IV TABLE 4: Type B(U), Type B(M) and Type C packages		APPENDIX V TABLE 5: Packages containing fissile nuclides, additional information		APPENDIX VI TABLE 6: Packages containing uranium hexafluoride, additional information	
Part 2		**Part 2**		**Part 2**		**Part 2**		**Part 2**		**Part 2**	
2.1.	Structural analysis	2.1.	Structural analysis	2.1.	Structural analysis	2.1.	Structural analysis	2.1.	Structural analysis	2.1.	Structural analysis
		2.2.	Thermal analysis	2.2.	Thermal analysis	2.2.	Thermal analysis	2.2.	Thermal analysis	2.2.	Thermal analysis
		2.3.	Containment design analysis	2.3.	Containment design analysis	2.3.	Containment design analysis	2.3.	Containment design analysis	2.3.	Containment design analysis
2.2.	Dose rate analysis	2.4.	Dose rate analysis	2.4.	Dose rate analysis	2.4.	Dose rate analysis				
		2.5.	Criticality safety analysis	2.5.	Criticality safety analysis	2.5.	Criticality safety analysis	2.4.	Criticality safety analysis		
						2.6.	Other analyses				

CONTRIBUTORS TO DRAFTING AND REVIEW

Boyle, R.   United States Department of Transportation, United States of America

Capadona, N.   International Atomic Energy Agency

Chrupek, T.   Nuclear Safety Authority, France

Dejean, C.   Institute for Radiological Protection and Nuclear Safety, France

Edwards, W.   World Nuclear Transport Institute

Elechosa, C.   Nuclear Regulatory Authority, Argentina

Ferran, G.   Nuclear Safety Authority, France

Fiaccabrino, V.   Nuclear Safety Authority, France

Fukuda, T.   Nuclear Regulation Authority, Japan

Garg, R.   Canadian Nuclear Safety Commission, Canada

Gauthier, F.   Institute for Radiological Protection and Nuclear Safety, France

Hirose, M.   Nuclear Regulation Authority, Japan

Katona, T.   Hungarian Atomic Energy Authority, Hungary

Komann, S.   Federal Institute for Materials Research and Testing, Germany

Konnai, A.   National Maritime Research Institute, Japan

Krochmaluk, J.   Institute for Radiological Protection and Nuclear Safety, France

Lizot, M.T.   Institute for Radiological Protection and Nuclear Safety, France

Malesys, P.   World Nuclear Transport Institute

Moutarde, M.   Institute for Radiological Protection and Nuclear Safety, France

Nöring, R.	World Nuclear Transport Institute
Pilecki, L.	Canadian Nuclear Safety Commission, Canada
Pstrak, D.	Nuclear Regulatory Commission, United States of America
Ramsay, J.	Canadian Nuclear Safety Commission, Canada
Reiche, I.	Federal Office for Radiation Protection, Germany
Sampson, M.	Nuclear Regulatory Commission, United States of America
Sarkar, S.	Australian Radiation Protection and Nuclear Safety Agency, Australia
Sert, G.	Institute for Radiological Protection and Nuclear Safety, France
Vaclav, J.	Nuclear Regulatory Authority of the Slovak Republic, Slovakia
Yagihashi, H.	Nuclear Regulation Authority, Japan

ORDERING LOCALLY

IAEA priced publications may be purchased from the sources listed below or from major local booksellers.

Orders for unpriced publications should be made directly to the IAEA. The contact details are given at the end of this list.

NORTH AMERICA

Bernan / Rowman & Littlefield
15250 NBN Way, Blue Ridge Summit, PA 17214, USA
Telephone: +1 800 462 6420 • Fax: +1 800 338 4550
Email: orders@rowman.com • Web site: www.rowman.com/bernan

REST OF WORLD

Please contact your preferred local supplier, or our lead distributor:

Eurospan Group
Gray's Inn House
127 Clerkenwell Road
London EC1R 5DB
United Kingdom

Trade orders and enquiries:
Telephone: +44 (0)176 760 4972 • Fax: +44 (0)176 760 1640
Email: eurospan@turpin-distribution.com

Individual orders:
www.eurospanbookstore.com/iaea

For further information:
Telephone: +44 (0)207 240 0856 • Fax: +44 (0)207 379 0609
Email: info@eurospangroup.com • Web site: www.eurospangroup.com

Orders for both priced and unpriced publications may be addressed directly to:
Marketing and Sales Unit
International Atomic Energy Agency
Vienna International Centre, PO Box 100, 1400 Vienna, Austria
Telephone: +43 1 2600 22529 or 22530 • Fax: +43 1 26007 22529
Email: sales.publications@iaea.org • Web site: www.iaea.org/publications

21-04014E-T